1

Wildflowers
of meadows and marshes

A Hamlyn Colour Guide

Wildflowers
of meadows and marshes

by Václav Větvička

Illustrated by
Zdenka Krejčová

Hamlyn
London • New York • Sydney • Toronto

Translated by Olga Kuthanová
Graphic design by Antonín Chmel
Designed and produced by Artia for
The Hamlyn Publishing Group Limited
London • New York • Sydney • Toronto
Astronaut House, Feltham, Middlesex, England

ISBN 0 600 35586 1
Printed in Czechoslovakia
3/15/06/51 — 01

CONTENTS

FROM WOODLAND TO OPEN COUNTRYSIDE

In large parts of central and north-western Europe the evolution of the climate following the last glaciation provided agreeable conditions for the development of the most highly organized type of vegetation — the forest — which is the characteristic feature of the landscape there. In the natural state this landscape would consist predominantly of a continuous, unbroken forest belt extending from lowland floodplain forests and oak woods to mountain spruce woods. In the warmest districts of this territory, however, there are places where post-glacial evolution probably took a different course than in the 'forest belt' region. These were the sites of the oldest agricultural settlements where the development of a closed forest cover was blocked by man — by ploughing and grazing.

Man, however, was not the only obstacle to the unimpeded development of a continuous, closed forest. In the rugged conditions of the alpine and subalpine belt treeless tracts are found with vegetation very similar to that of the period when a large part of the Continent was covered by the ice sheet. Finally, it is naturally difficult for a closed, continuous forest to be formed in exposed rocky situations and impossible in places permanently covered by stagnant or flowing water.

Forest communities, even though they may be viewed as the optimum type of plant cover, were relatively uniform. As the continuous forest cover of the late Bronze Age gave way to man's settlements and his activities what gradually emerged was a colourful mosaic of fields, meadows, pastures, ponds and human settlements interspersed with smaller or larger 'islands' of forest.

The landscape was so transformed that it now exhibits far greater diversity than the original virgin forests and includes many species of plants and animals that could not have flourished in the former, natural conditions. The present diversification of the landscape with man-made and substitute habitats enriched large expanses of this part of Europe with many organisms from so-called 'open' (i.e. unforested) landscape communities. It is only recently with the large-scale cultivation of crops that one can see again a simplification, a greater uniformity of the former colourful mosaic.

Within this complex of natural and secondary biotypes the vegetation of meadows, water and shores (the places where the two meet) stands apart in a special niche of its own.

MEADOWS – THE DIFFERENT KINDS

As many authorities agree, under natural conditions open meadows could have been formed only in the vicinity of the flood-zones of large water courses where the periodic passage of masses of ice following the spring thaw held the trees of the flood-plain forest at bay. Thus it is alongside Europe's rivers that we can most readily find natural or 'primeval' meadows.

However, many cultivated meadows were also located near water – not only in valleys and along river-banks but also by man-made reservoirs, ponds and lakes. This book identifies many of the characteristic meadow and freshwater plants in the north temperate zone.

Meadows as presented here are naturally not the only type of grasslands on the earth. The most closely related include mountain meadows, alpine grasslands, short-stemmed grasslands, pastureland (i.e. grassland that is short-stemmed due to grazing) and many xerophilous grasslands and steppes.

Extensive treeless, mostly grassy tracts are not limited to the temperate regions of the Old World. Similar tracts are to be found also in the tropics and subtropics – for example Africa's savannas and South America's pampas, campos and llanos. These are large grassy plains with only the occasional tree or group of trees. Most important of the climatic factors are the long periods of drought with periodic fires. That is why Africa's savannas, for example, have such a relative paucity of species – not every plant can withstand drought and fire! The prevailing grasses include millet (*Panicum, Pennisetum*), *Andropogon* and others. Often a single species of grass dominates whole tracts.

North America likewise has extensive grasslands called prairies, and it seems that here, too, frequent fires played a decisive role in their development by promoting the growth of grasses in competition with trees and shrubs. These grasslands, however, cannot be classed as meadows.

In practice a meadow is a piece of grassland whose grass is regularly mown for use as hay. Plants of a woody nature growing from seeds transported there or spread by vegetative means are intentionally removed. Most meadow plants, grasses in particular, are mesophytes, i.e. they are adapted to grow under conditions of well-balanced moisture, heat and food supply. Mesophytic plants also naturally embrace a relatively large group of organisms with many characteristics marking a transition to other types and other habitats – be it with respect to food, heat or moisture.

Man seeks causes for and rules governing all that occurs in nature,

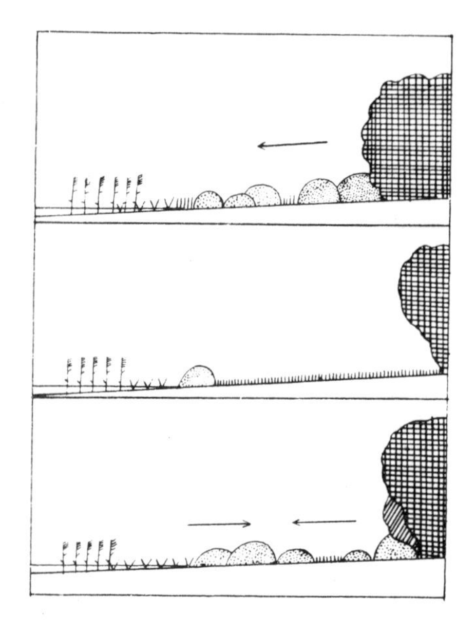

Land in the immediate vicinity of water was a place where the forest always had to struggle to retain a foothold. As a rule, however, these places were covered with softwood thickets, mostly willows, which spread from the forest margin to the water's edge (top). These places were eminently suitable for meadow cultivation. As long as these meadows were mown and tended by man the forest remained in its place and willows grew only by the waterside (centre). As soon as cultivation ceased spontaneous 'reafforestation' ensued — rapidly and in both directions (bottom).

in short he wants to have everything clear-cut and assigned to its own special niche. The same is true of botany. And so, scientists from the whole world joined forces and in 1966 came up with a comprehensive physiognomic-ecological classification of formations on the Earth. In this system terrestrial herbaceous formations, i.e. ones without tall woody plants, are classified as follows:

1. Meadows, pasturelands and related grasslands
2. Sedge marshes and spring-head formations
3. Steppes and related grasslands
4. Savannas and related grasslands
5. Herbaceous halophilous formations
6. Forb (any herb other than a grass) formations

Many plants of the first two groups (subclasses) will be introduced shortly, as will herbs of aquatic communities. In the latter system the plants are divided into the following subgroups:

1. Floating mats
2. Reed-beds
3. Communities of rooted, floating aquatics
4. Communities of rooted, submerged aquatics
5. Communities of free-floating, freshwater aquatics

A slightly different opinion was expressed almost half a century ago by Rübel, who called all herbaceous (non-woody) tracts 'meadows'. He made a distinction between:

1. land meadows, i.e.: a. steppes, xerophilous meadows
b. evergreen meadows
c. tall-stemmed meadows

2. marsh and water 'meadows', i.e.: a. marsh meadows (± shoreline)
b. submerged aquatic vegetation
c. moors

From both the examples above it is clear that the selection of meadow and aquatic plants in combination is no accident; these plant communities are very close and in some places even form a continuous mass of vegetation ranging from aquatic plants at the one end to land plants at the other. Determining the boundaries between the separate types and zones is extremely difficult because meadow and shoreline (marsh) plants include among their number many which are described as having a broad ecological amplitude, i.e. 'they grow everywhere'. Such plants are very tolerant of varying conditions in their environment because they have no specific requirements. One that serves as an example for all is the Common Reed (*Phragmites australis*) which grows in and by water margins, meadows and pasturelands, by waysides, and even in rock crevices. It thus tolerates not only complete immersion of its base in water but also extremely dry conditions.

As for meadow plants, in the stricter sense of the term, a suitable criterion for selection is the plant's relation to the water budget of the given habitat; they have been divided accordingly into the following groups.

Plants of dry and warm meadows are examples of meadow organisms with greater heat requirements that tolerate occasional dry periods or lower water table. Typical representatives are bromes

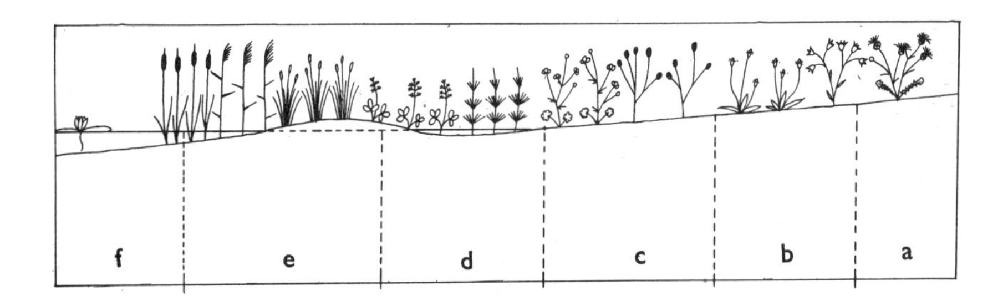

This simplified diagram illustrates not only the conditions in nature but also serves as a key to the groups of plants as presented in this book:

a. plants of dry and warm meadows
b. plants of dry and freshly-damp meadows
c. plants of damp to wet meadows
d. plants of bogs and moss-moor meadows
e. waterside and marsh plants
f. submerged and floating aquatics

(grasses of the genus *Bromus*) and such tracts could be called brome meadows. These meadows are usually mowed only once — as a rule in early summer — and then only cropped; they are generally found in hilly country, particularly on warmer slopes facing south or south-west.

Plants of dry and freshly-damp meadows may be considered classic examples of meadow vegetation. As everywhere here, too, grasses prevail (e.g. Oat-grass, Cock's-foot, Cat's-tail and Meadow-grass). However, inasmuch as this book strives to present a more colourful and variegated selection, this natural predominance is somewhat obscured by the many other meadow-flowers described which, though definitely not in the majority, are typical plants of meadows. From the viewpoint of plant sociology such meadows could be called oat-grass meadows. They are typical communities of rich mesophytic cultivated meadows from lowland to medium-high mountain elevations. The meadows are usually mown twice or more a year; in western Europe they are generally mown three times and then cropped in autumn. They are not dependent on the level of the ground-water table or occasional flooding.

A large and extraordinarily diversified group are the plants of damp to wet meadows. Depending on the prevailing species these can be referred to as fox-tail meadows, burnet meadows, meadow-sweet meadows, etc. These are usually meadows that are mown at least twice a year but often greater economic exploitation (mowing and drying of hay or follow-on grass) is prevented by water. Many plants of this group may at the same time be included in the following groups. One that will serve as an example for all is Meadow-sweet (*Filipendula ulmaria*): tracts of Meadow-sweet are a typical meadow community but the same plant also forms border communities alongside brooks and streams.

Places with stagnant water (high ground-water table) are conducive to the development of marsh-meadow communities, which almost always include Sphagnum and are thus sometimes referred to as moss-moor meadows. Soils in which plants of marsh and moss-moor meadows grow are usually deficient in nutrients (oligotrophic). Such meadows are unimportant from the economic viewpoint as well; the fodder, when it can be harvested during a temporary drop in the water level, is of inferior quality. Poor though they may be from the economic standpoint, such stands are rich and significant botanically for in such places one may encounter many plants that are rare and endangered species nowadays (for instance members of the orchid family). Other plants growing there — sundews — are again interesting for their mixotrophic, 'supplementary' diet. Alas, man's self-centered exploitation of every bit of soil is responsible for the fact that these lands are among the first to be drained. The change in the level and amount of water is immediately followed by a change in the composition of the vegetation. And so it is that the rarest marsh plants are becoming increasingly rarer and are threatened with extinction.

Our gradual shift from dry to damp conditions brings us to the junction of land and water, to the waterside. This is the place which is richest in plant and animal life. Hundreds and thousands of organisms, ranging from microscopic animals to tall, robust reeds, have found shelter and agreeable conditions for life and growth here. The remarkable number of organisms is matched by the diversity of species — and not just in a single place but on shores in general. The vegetation bordering a mountain stream is quite different from that along a lowland river whilst other varieties of plants grow on the shores of ponds and lakes and still others periodically appear on the exposed bottoms.

Most waterside and marsh plants are fond of sun and light. There is no reason for them to fear even extreme heat, for they do not need to economize on water, having plenty on hand all the time! They grow rooted in mud or in periodically flooded soil; many are submerged with only a small part above water. Reed-beds, meaning stands including not only the Common Reed but also sweet-grass, bulrush, reedmace, and the like, are typical vegetation of this sort. They are generally found in the shallow margins of ponds and lakes and are distinguished by profuse and rapid spreading by vegetative means. In shallow bodies of water they may even be very troublesome weeds covering vast expanses within a short time. Not only do these masses of vegetation increase the rate of evaporation of the water, they also impede or even block navigation and irrigation systems, and damage pond husbandry. Reedmaces and reeds, for instance, covered about 90 hectares of inundated fields in the River Dyje valley, in Czechoslovakia, within a period of three years. Their rapid spread, however, can be put to good use, for example in city sewage water clarifying tanks, for reeds accumulate mineral elements in their tissues and may thus serve as biological filters.

The extent of a reed-bed may be increased as much as three-fold within a single growing season. This prevailing method of multiplication by vegetative means often leads to their forming vast monotonous communities or mosaic-like masses with the various individual species forming separate islands. Another characteristic feature of these communities is a certain zonality — border growth by the waterside influenced as a rule by the level of the water and flooding.

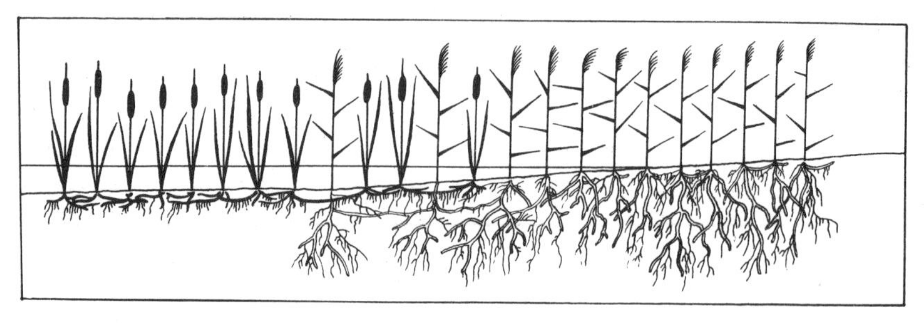

The margins of bodies of water are often overgrown with reeds and reedmace. The common reed grows closer to the shore (in shallower water) and its root system penetrates to greater depths. Reedmace, on the other hand, spreads into deeper water and farther from the shore; its root system, however, is relatively shallow.

Reed-beds and adjoining stands of large sedges are very difficult to keep in check and contribute to the silting up of bodies of water. The amount of organic matter (biomass) represented by reed-beds is enormous and their accumulating remnants speed up the formation of mud; the dense tangle of their rhizomes and underground runners further fortifies this mud — all of which reduces the area and volume of the body of water. As determined some years ago in several places in South Bohemia, Czechoslovakia, on every square metre of shore there were 50 to 80 specimens (i.e. stems) of reedmace or 80 reed stems, 370 bulrush stems, 130 sweet-grass stems or 65 bur-reed stems. Also worthy of note is the weight of this organic matter: fresh stems gathered on a single square metre weighed 8 to 10 kg in the case of reedmace, 12 kg for sweet flag, 3.5 to 5 kg for reeds, 6.5 kg for bulrush and about 6 kg for sweet-grass. The productivity of living matter, as can be seen, is extraordinary and not only has it been exploited by man from time immemorial — as bedding for stock, roofing, construction and insulating material, as a source of cellulose, and to weave baskets, matting, and the like — but new uses for it are continually being sought.

Whereas waterside and marsh plants tolerate an excess (and occasionally also a scarcity) of water, aquatic plants are practically unable to do without. They require it in abundance for normal growth, particularly those plants that are not rooted in the soil at the bottom of still or flowing water. Water plants may be entirely submerged or they may have most of their leaves floating on the surface, or the whole plant may be floating freely; the first two are rooted in the bottom, the last have trailing roots (though some may have none at all).

It would be a mistake to believe that an aquatic environment does not provide as good conditions as land for the evolution and development of plant communities. True, in the north temperate zone communities of floating aquatics are relatively poor in terms of diversity of species. Mostly they are duckweed and frogbit communities. The first could be described as communities of periodic occurrence and duration. Even the occurrence of individual species of duckweed is not constant during a single growth period and the composition of such a community changes during the course of the year. Frogbit communities, on the other hand, are relatively stable or at least long-term, that is if the conditions of the aquatic environment remain unchanged. However, changes such as marked fluctuations in the water level, or even occasional short-term drainage, do not have any pronounced effect on communities of water plants rooted in the bottom. Nevertheless, even they are not as rich in species as are meadow communities. As a rule the prevailing plants are various pondweeds, White Water-lilies and Yellow Water-lilies. Another adaptation to occasional draining of ponds and fluctuations in the water level is that these plants develop land (terrestrial) forms sometimes quite different in appearance from the normal submerged plant. Examples are Amphibious Bistort, Fringed Water-lily and Broad-leaved Pondweed.

Every aquarist will confirm the importance of submerged aquatics for aquariums. Such plants greatly influence the properties of the water (amount of oxygen and carbon dioxide, acidity, and the like). Many water plants serve as an indication of the water's purity. Fishermen and pisciculturists view submerged aquatics as a welcome element; masses of such vegetation provide fish with temporary shelter and are ideal places for spawning. Only large-leaved floating plants such as the Yellow Water-lily and excessive numbers of certain duckweeds in nutrient-rich (over-fertilized) bodies of water have an adverse effect in that they cut down light.

Submerged aquatics and marginal plants of shallow water also shelter many other living organisms. First and foremost they are suitable, firm objects on which aquatic animals climb, lay their eggs and hide. Observations of 80 pondweed leaves in summer disclosed that during that period they were populated by about 300 protozoans, 10 rotifers, 40 other worms, 137 bryozoans, 17 molluscs, 500 minute crustaceans, 70 may-fly larvae, 22 mites, 3 butterfly larvae, 1 caddis fly, about 800 midges and 454 aphids. Even more impressive was the count on the leaf-rosettes of eight Water-chestnuts which totalled more than 16,000 worms, 9,000 small crustaceans and almost 6,500 midge larvae!

Waterside plants and reed-beds provide shelter for many species of water birds. Ducks, geese, grebes, bitterns, coots and other birds build their nests of reed and grass stems here, as do many warblers whose ceaseless song breaks the stillness of lakes and ponds.

Grouped together with water and marsh plants but set apart is the vegetation on the bottoms of exposed and summer-drained ponds. It is composed of plant communities distinguished by rapid but short-term growth. The initial stages are evident soon after the ponds are drained. Plants of exposed bottoms generally produce large amounts of seeds that retain their powers of germination for a long time. They are conspicuously photophilic and do not tolerate the competition of tall herbaceous plants. Their opportunity arrives with the appearance of new 'virgin' soil. Within a very short period they germinate, grow, flower, produce mature seed — and the pond can be filled again. If it were not, the vegetation would continue developing into forms of terrestrial vegetation made possible by the properties of the exposed bottom. With the appearance of tall marsh and meadow plants, however, the initial vegetation of the exposed bottom would not be able to survive the competition and would be stamped out one way or another.

Typical plants of exposed bottoms are annual herbs, but often found there as well are terrestrial forms of certain typically aquatic plants such as Water Crowfoot, *Polygonum,* pondweeds and others. Also frequently encountered in the poor sandy soil of exposed bottoms are plants of abnormal, dwarf habit, with only a few leaves, small stem and scanty inflorescence.

The growth and spread of the plant cover of an exposed bottom sometimes resembles an explosion. The bottom turns green in practically one day and within a few weeks this vegetation disappears with equal suddenness. Where the seeds of these plants come from remains one of nature's 'puzzles'. Sometimes there isn't a single pond or lake with an exposed bottom in the surrounding countryside for several years and yet at the very first opportunity they appear over and over again.

COLOUR ILLUSTRATIONS

Plants are arranged in groups according to the dampness of their environment and within each group according to their flowering period.

Cypress Spurge
Euphorbia cyparissias L.

Euphorbiaceae

The order of spurges (Euphorbiales) comprises a group of isolated families whose mutual relationships are difficult to establish. This is doubtless because many have a very specialized metabolism – involving caoutchouc (raw rubber) and poisonous substances of a protein character – and, for example, in members of the genus *Euphorbia,* an unusual flower structure. The spurge family (Euphorbiaceae) includes many popular house plants (*Croton, Poinsettia*) as well as plants of economic importance, like *Hevea* (rubber tree) and *Ricinus* (castor-oil plant). Most members of the spurge family grow in tropical regions. Only the genera *Mercurialis* and *Euphorbia* are also found in temperate regions.

Spurges contain lactiferous ducts that secrete latex, a milky liquid which oozes from the plant when it is bruised or cut. If you pluck a Cypress Spurge on a dry sunny bank you can make bubbles by blowing through the stem, a popular pastime of village children in bygone days.

Cypress Spurge is often attacked by rust (*Uromyces pisi* Pers.) which causes deformation of the plant's classic structure: the long stems bear only shortly-oval leaves coloured a bright orange on the underside and smelling of 'honey' (1).

Cypress Spurge is one of 1,600 species of spurges found throughout the world with the exception of the arctic regions. It is a perennial with underground rhizome and erect, tufted stem. The upper stem leaves are narrowly linear. They resemble the needles of conifers and that is why this spurge was given the Latin name *cyparissias,* meaning like a cypress or, in the wider sense, coniferous.

The striking shape of the flower is known as a cyathium — a terminal umbellate inflorescence subtended by bracts. In the axils of the two opposite bracteoles rests a bisexual inflorescence with a small central ovary on a long stalk and in the axils of five bracteoles are small, jointed stamens; in between the bracts are orbicular, two-horned glands. The fruit is a warty capsule with three seeds. The flowering period is from April to July.

Upright Brome

Bromus erectus Huds.

Gramineae

In this book Upright Brome is the only representative of the genus *Bromus* which embraces approximately 50 species all told. Some have become troublesome weeds, e.g. Rye Brome (*B. secalinus* L.) and Field Brome (*B. arvensis* L.). Brome is highly resistant to drought and is therefore often used in grass mixtures for railway embankments and roadsides. In the wild it occurs from lowland to mountain elevations (up to 1,700 m above sea level), being partial to lime-rich soil. It is both an ameliorative grass (one that binds the soil of newly-made embankments with its roots) and a forage grass, but is of mediocre quality and median or low yield (after the first mowing its growth is slow and scant). However, as it has its uses in grass mixtures, it has spread beyond the limits of its original range (central and southern Europe, north Africa, Asia Minor).

In forestry Upright Brome is known as one of the several initial grasses to cover new clearings. Its sudden presence may also be a warning signal, the first sign of the drying up (and deterioration) of what was once fresh, nourishing soil.

Bromes are grasses with closed leaf sheaths and panicle-like inflorescences of multiflowered spikelets. The generic name *Bromus* is derived from the Greek — either from the word *brómos*, the ancient Greek word for the similar oat *(Avena sativa)* or Wild Oat *(A. fatua)*, or from the word *bróma*, meaning food (probably in terms of cereal).

Upright Brome is a plural-yearly herb with tufted rhizome covered with brown, fibrous sheaths. The young leaves are inrolled, the leaf sheaths tubular nearly to the base of the blade. The ligule is strikingly small — barely 1 mm high. The margin of the leaf blade is covered with long hairs; the sheath is also hairy.

The culms are rigid and about 1 m high. The inflorescence is a narrow, contracted panicle; the spikelets forming the panicle are composed of 3 to 12 flowers (1).

The flowering period is in June and July.

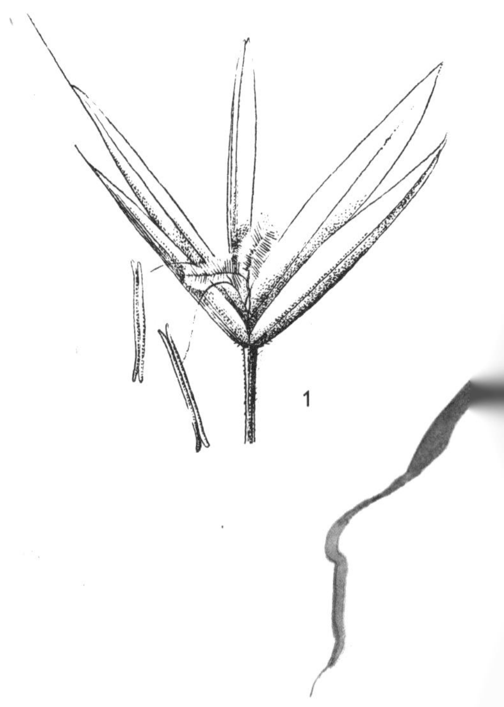

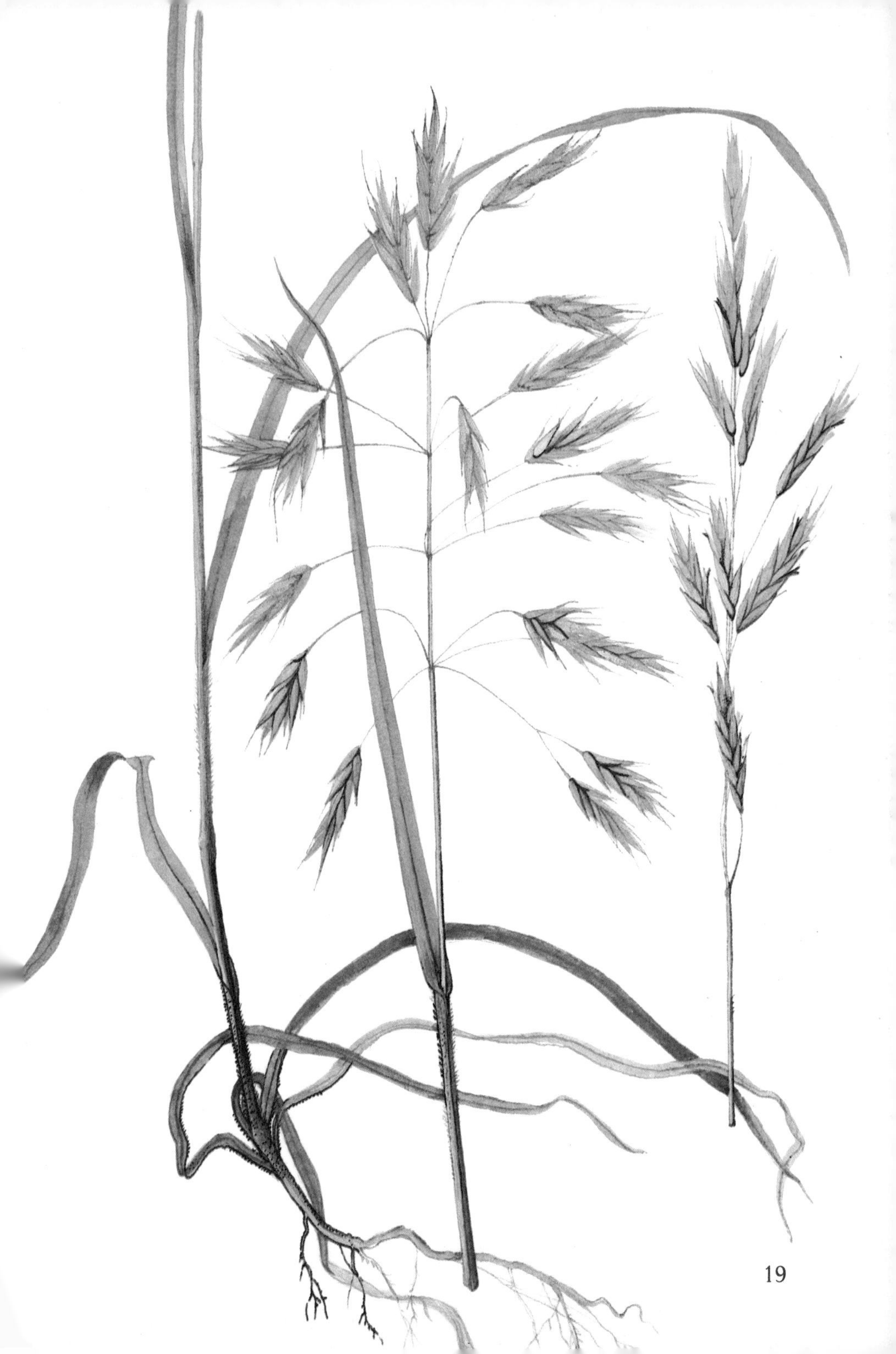

Common Crown-vetch Fabaceae
Coronilla varia L.

The Mediterranean region, Canary Islands and Near East are the native lands of legumes of the genus *Coronilla.* From there they spread, either by natural means or through the agency of man, to districts farther north. Nowadays some may be encountered in central and north-western Europe and, as for the Common Crown-vetch, it is even believed to be indigenous to these parts. It likes warm, dry, sunny places such as hedgerows, dry meadows and pastureland. It may even be found at higher elevations in places with calcareous soil — for instance in the Sudeten Mountains, Alps and Jura Mountains. Its spread here was probably aided also by man; by deforestation of south-facing mountain slopes, long-term grazing by livestock and cultivation of meadows.

The roots of Common Crown-vetch, like those of other plants of the pea family, are covered with irregular bumps and tubers. These swollen tissues are inhabited by bacteria that invaded the body of the host plant from the soil. These bacteria have the ability to fix atmospheric nitrogen and that is why plants of the pea family are also used in agriculture as forecrops, prior to planting crops that require nitrogen-rich soil.

Some members of the pea family, including Common Crown-vetch, are noted for the so-called sleep movements of the leaves — the separate leaflets of the compound leaves are pressed together (upward) for the night. This, of course, has nothing in common with the sleep of animals. It is merely the plant's reaction to the amount of incidental light and is explained by a change in the turgor of the tissues at the point of bending.

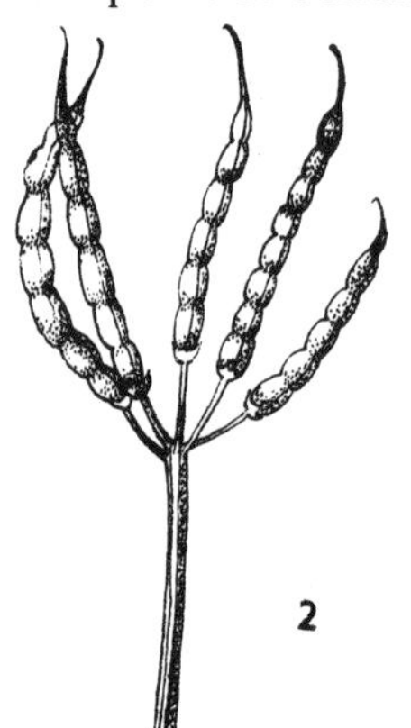

Vetches are annual or perennial herbs (some Mediterranean vetches are even shrubs) with alternate leaves composed of 1 to 12 pairs of small leaflets. The flowers, like all flowers of the pea family, consist of a bell-like calyx with short teeth and a corolla composed of a round, clawed standard (large upper petal), ovate wings (two lateral petals) and keel (two fused lower petals). One of the stamens is free, nine are fused. Most vetches are yellow but the Common Crown-vetch is an exception — it is coloured white-pink and violet. The whole inflorescence resembles a small royal crown and hence its Latin name *Coronilla,* meaning small crown. The fruit of this vetch is an upcurved loment with constrictions between the seeds (1) that splits at maturity.

The flowering period is in June and July.

Burnet Saxifrage
Pimpinella saxifraga L.

Umbelliferae

Plants of the parsley family are noted for their aroma and Burnet Saxifrage is no exception. In days of old it was the custom to add Burnet Saxifrage to beer, and even to mull wine with it! Country folk knew the value of newly-greening meadow herbs. In former times when vegetables we consider common were not available the spring soup made throughout practically all of Europe was sure to include the tender young leaves of Nettle, Dead-nettle, Caraway, Chicory, and Burnet Saxifrage. A few weeks later the leaves of the same plants, larger but still young, were used to make a tasty spring salad.

Burnet Saxifrage grows throughout all Europe in dry meadows in the lowlands as well as higher mountain ranges (in the Sudeten Mountains, for example, up to elevations of 1,300 m). In the Giant (Krkonoše) Mountains, in the important, species-rich locality of Sněžné Jámy, there is even an endemic strain — *P. saxifraga* subsp. *rupestris* Weide. The related Greater Burnet Saxifrage (*P. major* (L.) Huds.) is also an important European meadow plant. Both are likewise important drug plants: the part collected for medicinal purposes (in spring or autumn) is the root, used as an expectorant particularly in the treatment of asthma and catarrh of the respiratory passages. Burnet Saxifrage is a plant that was believed to be endowed with magic powers in medieval times — it was used to ward off the plague.

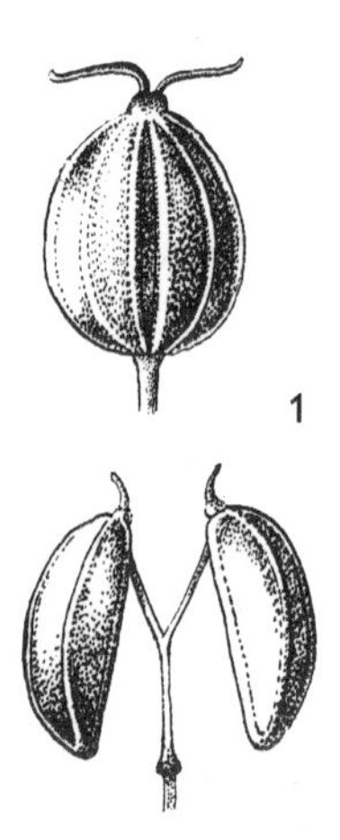

Pimpinellas are annual, like Anise *(P. anisum),* or perennial herbs with pronounced heterophylly (bearing leaves of different forms on the same stem). The basal leaves of Burnet Saxifrage are odd-pinnate with rounded, deeply-toothed leaflets, the stem leaves are divided into narrow, linear segments. The umbels (inflorescence typical of the whole family) are composed of many small umbellets and those in turn of tiny hermaphroditic or only staminate (male) flowers. The flowers are creamy-white, very occasionally pink.

Greater Burnet Saxifrage (*P. major* (L.) Huds.) has a sharply angled, hollow stem and leaves mostly of the same shape; Burnet Saxifrage has a rounded, finely-grooved stem that is usually solid.

Pimpinellas contain up to 0.4 per cent of essential oils, furocoumarins (with photosensitizing effect) and many other substances, including resins and sugars. The fruit is a double achene (1).

The flowering period is from June to September.

Hoary Plantain
Plantago media L.

Anyone who has ever tried to establish an 'English' lawn on the Continent will have been caught unprepared, in about five years, by the appearance of numerous plants with broad leaves arranged in a basal rosette: plantains — usually Hoary Plantain and often also Great Plantain (*P. major* L.). Both species are indigenous plants of meadows, pastures and open hedgerows — Great Plantain grows in moister places, Hoary Plantain in drier situations. Great Plantain is also a typical plant of trampled roadsides and places like waste dumps.

The plantains are the only family of the order Plantaginales and the genera that make up the family are also few in number. As for the genus *Plantago* itself, it embraces more than 260 species with an extremely wide ecological amplitude. Besides the mesophytic plants of Europe's meadows they include definitely halophilous plants such as Sea Plantain (*P. maritima* L.) and rock-garden favourites such as *P. nivalis* Boiss. from the mountains of sunny Spain, which looks more like a small tuft of cotton than an ordinary plantain.

Plantains have striking ground rosettes of leaves with characteristic venation (the principal veins are parallel) resembling that of Monocotyledons. These veins (vascular bundles) have spiral stiffeners — if you try to tear the leaf of Hoary Plantain in half crosswise the two parts of the blade usually remain joined by threads of stiffened vessels.

The sessile, 4-merous flowers of plantain are arranged in spikes. Hoary Plantain (1) has scapes that are always longer than the flower spike and leaves that are practically stalkless. The corolla lobes are silvery white. Great Plantain has brownish corolla lobes, leaf stalks the same length as the blade (2) and, first and foremost, a conspicuously long spike (3) — usually longer than the scape.

The flowering period is from June to September.

1
3

Greater Knapweed
Centaurea scabiosa L.

Compositae

If you wanted to find Greater Knapweed in medieval herbals you would have had to look for it under the name of St. Zacharias' plant (*Herba St. Zachariae*). How the two — the *Centaurea* and St. Zacharias — came to be linked, however, is something we will never learn.

Greater Knapweed is a prominent meadow plant of xerophilous and thermophilous meadow communities, sunny and bushy banks, hedgerows, fallow land and field margins found from the lowest elevations to the foot of mountain ranges. However, it is also a very variable plant. The most striking and conspicuous characteristic is the structure of the involucral bracts in the inflorescence. The diverse and apparently hereditarily constant variability of this character naturally led to the plant's differentiation into a great number of intraspecific taxons (mostly varieties, subspecies) bound to small territories.

The striking comb-like appendages on the bracts of the Greater Knapweed are also an instance of the occurrence of the colour 'black' in the plant realm, caused solely by the concentration and combination of brown pigments and anthocyanins.

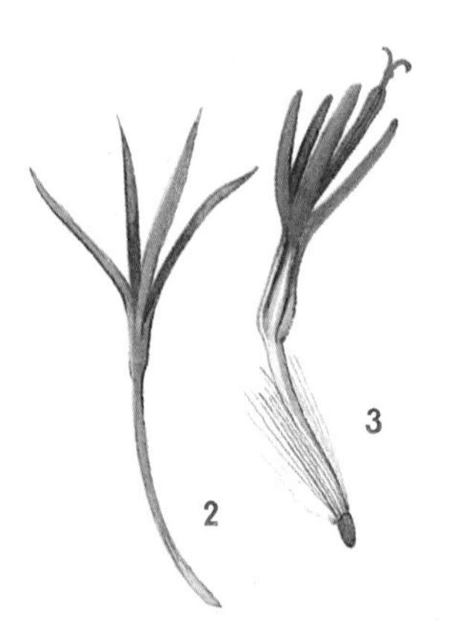

There are some 700 species of *Centaurea* in the northern hemisphere. The Greater Knapweed is one of the most variable of the lot. It is a 30—150-cm-high herb with a rhizome that becomes woody (1) and irregularly pinnatifid leaves with entire or pinnatifid lobes; the basal leaves are usually markedly different from the upper stem leaves.

The flowers are generally reddish-violet, only very occasionally pink or white. The marginal florets are usually sterile, strikingly large and radiate, and serve as visual bait for insects (2). The florets in the centre of the inflorescence are hermaphroditic and fertile (3). The fruit is a 4—5-m-long achene (cypsela) with pappus (a double-rowed tuft of short hairs).

The flowering period is from July to September.

1

Common Agrimony
Agrimonia eupatoria L.

Rosaceae

Agrimony, highly regarded and valued as a medicinal plant since ancient times, may be found on practically every dry, sunny bank, in dry and warm meadows, thickets and hedgerows from lowland to mountain districts.

It was used in the form of an infusion to treat various internal diseases — hepatitis and other diseases of the liver, digestive disorders, intestinal ailments and painful joints. Externally it was used in the form of a bath or douche and the washed leaves or top parts were even applied as compresses directly to wounded skin. Modern medicine also finds uses for it to aid and promote digestion, to promote the flow of bile and as an astringent in cases of diarrhoea. It is also used externally as a mouth gargle. The plant's healing properties are also reflected in several of the common names by which it is known in German-speaking districts, e.g. *Leberkraut* (meaning liver herb).

Agrimony is also a remarkable plant from the morphological aspect: the basal parts of the flower are joined into an obconical, cuplike hypanthium edged with deflexed bristles. These bristles, soft in the bud and flower, become hard (woody) in late summer and catch in the fur of animals and garments of passers-by, thereby aiding in the dispersal of the species.

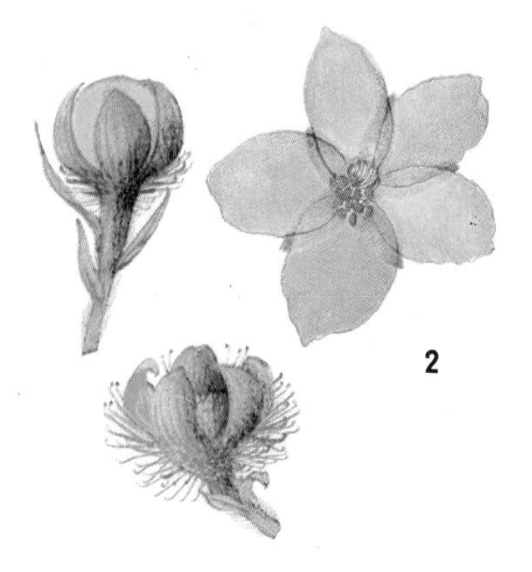

Common Agrimony is one of 10 species of agrimony. It is a hairy, glandular perennial herb with thick rhizome and little-branched stem. The compound leaves have a striking shape composed of alternating pairs of larger and smaller leaflets (1). Such leaves are said to be odd-pinnate with interjected leaflets. The yellow flowers are five-petalled (2), with 5 to 20 stamens, and regular; they remain open three days at the most, after which the petals fade and fall.

The top parts contain up to 5 per cent tannins, essential oils, bitter principles and other substances. They are collected for medicinal purposes from July to August, but only with stems less than 0.5 cm thick.

The flowering period is from June to August.

Wild Daffodil

Amaryllidaceae

Narcissus pseudonarcissus L.

At the turn of the 19th century (and in some places perhaps even several decades later) vast meadows of narcissi could still be seen in many parts of the Alps and elsewhere in western Europe. Photographs from that time (e.g. from the neighbourhood of Avants in Switzerland) show meadows smothered with narcissi. Alas, there are few such places nowadays — narcissi have succumbed to the inroads of civilization. The east-Carpathian populations are somewhat better off — one of the Transcaucasian valleys has been named the Valley of Narcissi and proclaimed a rigidly protected nature preserve.

The centre of distribution of the west-European meadow narcissi included the Iberian Peninsula, France, England, Belgium and Italy. Growing wild there were the yellow Wild Daffodil and the white Pheasant's Eye Daffodil (*N. poëticus* L.) and the related *N. exsertus* (Haw.) Pugsl.

The flood of fragrant narcissi in spring is apparently what inspired the traditional May *Narzissenfest* festivities in German regions. The great variety of species, variability and ease of cross-breeding were put to good use in the flower trade: for instance in the middle of the 20th century England alone had more than 8,000 registered varieties of large-flowered narcissi. The various forms differ not only in colour (ranging from white to orange) but also in colour combinations and structure of the flower, particularly regarding the corona which can be very short or on the other hand long and trumpet-like. The many varieties also include multiflowered as well as double forms.

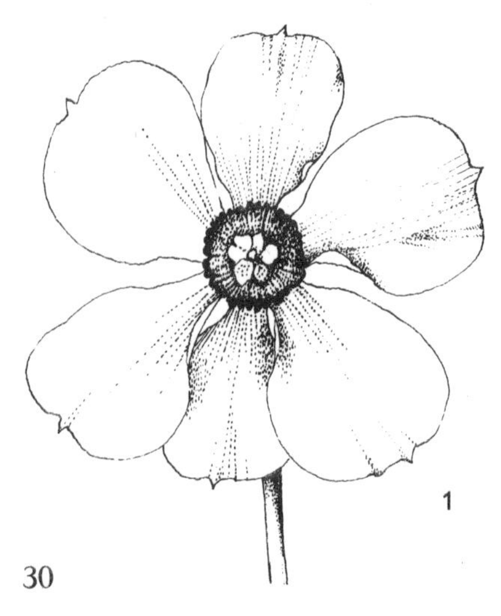

The narcissus takes its name from the Narcissus of Greek mythology; it may also be derived from the Greek word *narkáo*, meaning to stupefy or dull the senses — probably because of the flower's penetrating fragrance or poisonous properties.

Narcissi are perennial herbs with underground bulb and long, erect, linear leaves coloured greyish-green. The scape is round (rarely angled) and hollow. It is usually terminated by a single flower, though there may sometimes be several,

growing from the axil of a membranous bract. The flowers, which are hermaphroditic, are usually nodding, pendent and regular with six, generally spreading perianth segments and a corona inside the corolla. Those of cultivated varieties (1) often have a corona and corolla of contrasting colour.

Narcissi flower from March until the end of May, but like many bulbous plants they can be made to bloom much sooner if cultivated.

Daisy
Bellis perennis L.

Compositae

This herb, that grows 'by every wayside', is one of the smallest and daintiest of nature's flowers. As a matter of fact not only the Daisy but the whole genus *Bellis* has an extensive distribution embracing Europe, America, non-tropical Australia and New Zealand. Ornamental large-flowered or double forms, usually of *Bellis perennis*, are grown in gardens throughout the world.

Its smallness and absence of any special growing requirements predetermined its being viewed as a symbol of humbleness; it also figured in many legends, from ancient Greece to Norway.

From the biological aspect daisies are of interest for their hardiness to cold (they can tolerate even temperatures of −30 °C) and for their reaction to mechanical irritation: when touched the marginal ray florets rise. The flower-heads were also used in experiments for the biometric evaluation of the variability of species.

The Daisy was also used in folk medicine. An infusion from the dried flowers was recommended in the treatment of lung diseases, catarrh of the upper respiratory passages and, externally, for treating skin diseases.

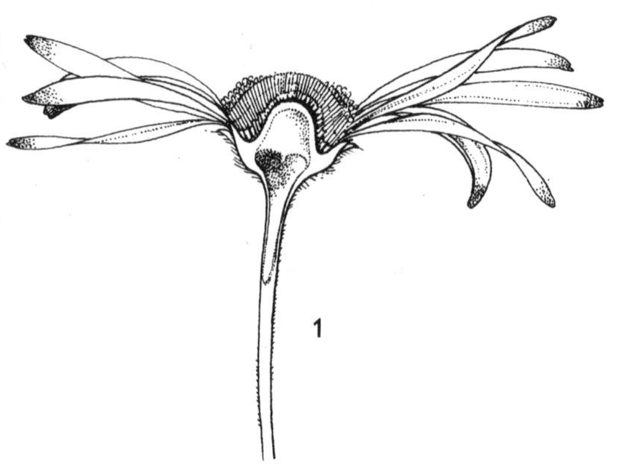

The Daisy is a common component of fresh, nutrient-rich meadows, pastures and parkland turf. It is a 5—15-cm-high perennial herb with leaves forming a ground rosette. The scape is usually terminated by a single inflorescence, a so-called 'biological flower' (1). The flower-heads of wild daisies measure barely 15 mm across, but cultivated, double varieties include 'giants' several centimetres in diameter. The yellow disc flowers are tubular and hermaphroditic, the ray flowers are ligulate, white or pink, and usually unisexual or sterile.

Though the flowering period is generally listed as being from March to November, these lovely little daisies may be encountered practically all year round — that is if they are not covered by snow.

Field Scabious

Knautia arvensis (L.) Coult.

Dipsacaceae

Viewed from above Field Scabious looks like some *Centaurea* or similar composite flower; the flower-head definitely resembles those of the composite family. Within the order Rubiales (madder), however, Marsh Valerian (*Valeriana dioica*), for example, is much more closely related to Field Scabious than any *Centaurea!*

Field Scabious was formerly classed in the genus *Scabiosa* but besides other characteristics the two – *Scabiosa* and *Knautia* – can be distinguished by the shape of the corolla: in members of the genus *Scabiosa* it is five-lobed whereas in members of the genus *Knautia* it has only four lobes. Field Scabious is an extremely variable plant and often occurs in meadows with flowers coloured creamy-white or pink instead of bluish-lilac; it is, therefore, not surprising that in drier situations where it may grow side by side with the creamy-yellow Yellow Scabious (*Scabiosa ochroleuca* L.) people often mistake the one for the other.

Field Scabious is native to the temperate regions of the Old World where it grows in fresh, damp or drier meadows, ditches, by field paths and occasionally also in border thickets.

Field Scabious is a 30–150-cm-high perennial herb covered with spreading hairs. It has an underground rhizome and upright stems. The basal leaves are usually undivided, the upper stem leaves deeply cleft and the uppermost pairs of opposite leaves are deeply lyrate to pinnatisect. This conspicuous heterophylly (growing leaves of different forms on the same stem) is the characteristic that exhibits the greatest variability.

The flowers are arranged in heads (capitula), which may be composed of hermaphroditic flowers or only male flowers. Heads of male florets are always smaller. The individual lilac flowers (1) are conspicuously zygomorphic.

Practically every population includes specimens with flowers of different hues, often even pink (2). Also very attractive are heads of flowers that are still in the bud (3). The fruit is a one-seeded achene.

Field Scabious flowers from May onward, sometimes until September.

Meadow Saxifrage
Saxifragaceae

Saxifraga granulata L.

Meadow Saxifrage is a typical meadow plant of western and central Europe generally found in sandy, mostly lime-deficient soils, in rather dry hedgerows, on banks and in pastureland. In Europe it occurs as several strains: the typical subspecies (subsp. *granulata*) has a continuous distribution throughout the whole of central and north-western Europe, excepting northern Scotland, the coast of Holland and the like; in Ireland and Scandinavia it occurs only occasionally and in the Alps it is absent altogether — however, it also grows in Portugal and north-west Africa.

The most noteworthy aspect of its biology is its unusual method of vegetative propagation. Growing in the axils of the basal leaves are small bulbils that can separate from the parent plant and grow into new plants.

At one time the top parts of Meadow Saxifrage were used in folk medicine to treat skin diseases and as a diuretic. In ancient times it was used in the treatment of kidney stones and gall stones — hence, supposedly, the name *Saxifraga* from the Latin words *saxum,* meaning rock, and *frangere,* meaning to break.

Meadow Saxifrage is a little-known herb. Every gardener, however, knows the many saxifrage varieties suitable for the rock garden; rock crevices in high mountains is where they grow in the greatest numbers. That, perhaps, is the more likely origin of their name.

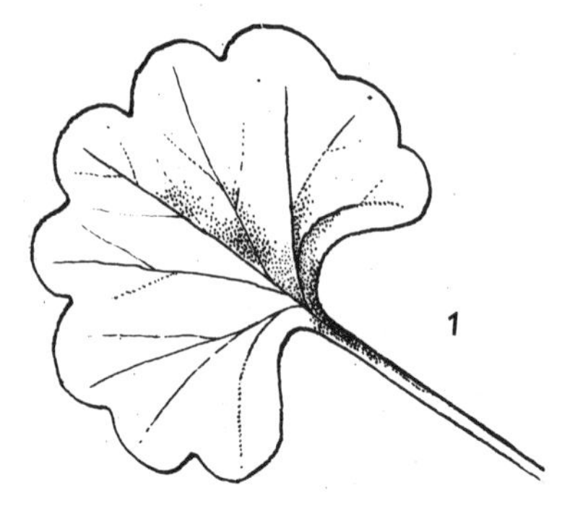

Of the approximately 310 species of saxifrage, Meadow Saxifrage is one of the few lowland, meadow species. It is a loosely-tufted perennial herb with basal rosette of rounded-kidney-shaped leaves (1) with crenate margin. The upper stem leaves are sessile, elongate and lobed at the tip. Bulbils are formed only at the base of the basal leaves, never in the axils of the stem leaves, as is the case in the

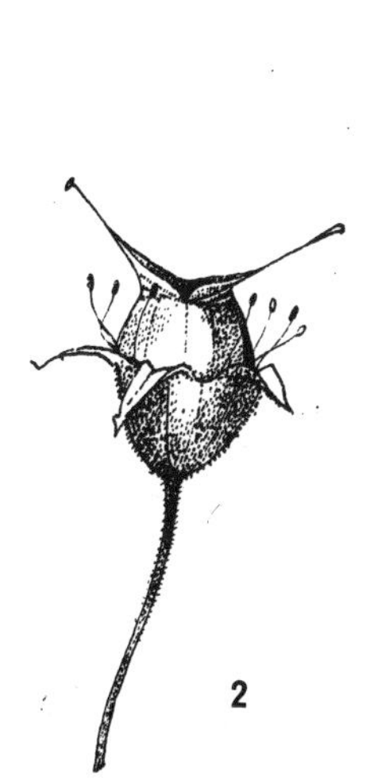

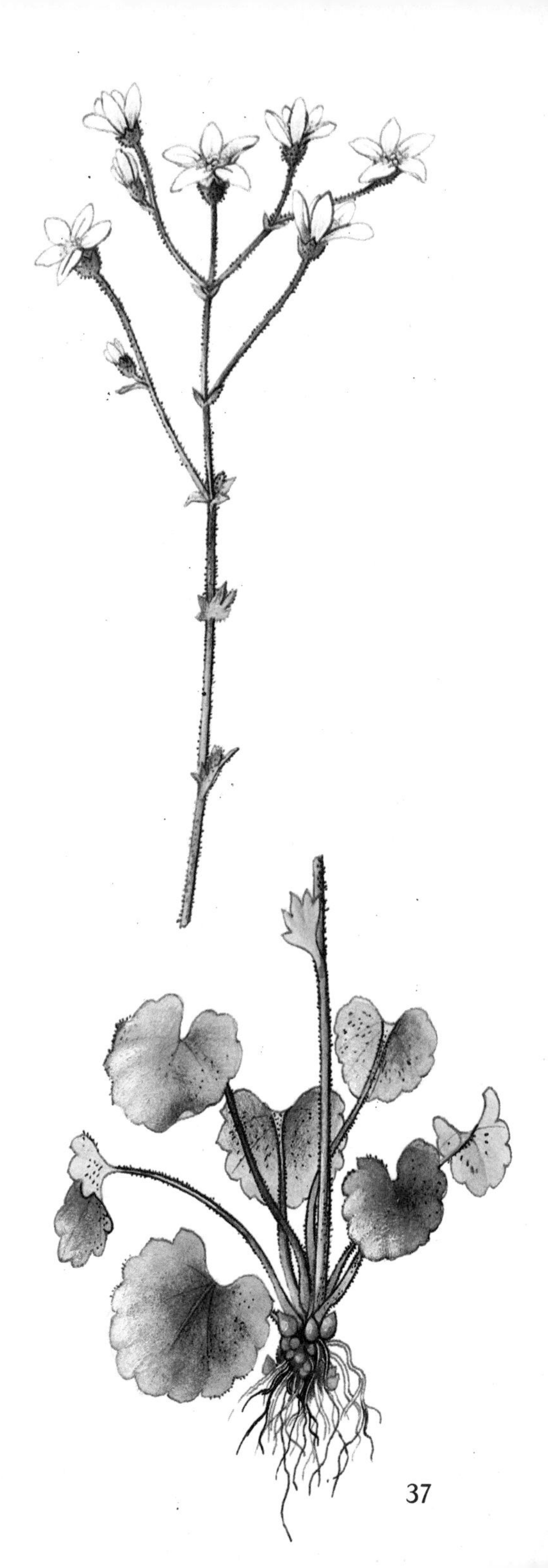

related Bulbous Saxifrage (*S. bulbifera* L.). The top parts contain tannins, bitter principles and the glycoside bergenin. The stem is loosely branched and bears only a few flowers which are white and glandular. The fruit is a broadly ovate capsule (2) containing tiny, light seeds without storage tissues.

The flowering period is from May to July.

Cock's-foot

Gramineae

Dactylis glomerata L.

Cock's-foot is a very adaptable grass found in both fresh and drier meadows. It is also a grass of great economic importance. Though it grows even in poor soil, it does best in soils that are mesotrophic to highly eutrophic (rich in nutrients). That, also, is where its great competitive power is most manifest — it was proved that in cultivated grasslands it is able to suppress such weeds as Coltsfoot and Dandelion. Inasmuch as this competitive power may also have undesirable effects by suppressing other valuable grasses — in ten-year-old spreads Cock's-foot already accounts for 50 per cent of all the grasses, particularly in early spring — Cock's-foot has become the subject of research by breeders aimed at finding later-flowering strains. In western Europe alone cultivated varieties of Cock's-foot represent 12 per cent of all cultivated grasses — in other words more than any other. Great Britain grows eight cultivated varieties, the German Federal Republic 14, and Denmark 10.

Cock's-foot has also been studied by geneticists. European populations include plants with different chromosome numbers (sometimes described also as separate taxons) and this fact has been made use of in the selection of cultivated varieties.

Cock's-foot belongs to the group of grasses with two- to multi-flowered spikelets and two glumes (1), a group that is very old from the evolutionary standpoint. The scientific name *Dactylis* is derived from the Greek word *dáctylos*, meaning finger, for the open inflorescence resembles a hand with fingers.

It is a densely-tufted, plural-yearly grass with a robust root system. It is readily identified, among other things, by the fact that the leaf sheaths and base of the culms are conspicuously compressed (laterally flattened) (2) with sharp edges. The leaf blade, with toothed margin, is rough. The ligule is membranous, up to 4 mm long, tapering to a point and often fringed (3).

The flowering period is from May to July.

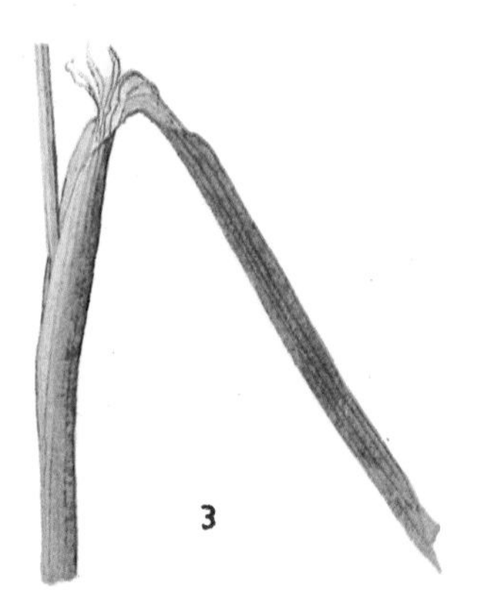

1
2

Common or Smooth Meadow-grass

Gramineae

Poa pratensis L.

Meadow-grasses are typical representatives of the large family of grasses. Common Meadow-grass is an important hay and pasture grass and a permanent component of long-term meadows and pasturelands because it provides good-quality fodder, particularly in the first half of the growing season (in the second half the yield of green fodder is lower). Common Meadow-grass is noted for its rapid spread and thus quickly occupies empty places. It is often used in lawn mixtures.

In the wild it grows in rather dry, nourishing soils but may also be found in forest margins, thickets, and by waysides from lowland districts to elevations of more than 2,000 m. It is distributed throughout all Europe (as far as Spitzbergen in the north), in northern Asia, North America and Australia.

A common component of natural, freshly-damp and nutrient-rich meadows in northern and central Europe is the related Rough Meadow-grass *(P. trivialis* L.*)*, with long creeping stolons, which is usually used in special grass mixtures for playing fields.

The broad ecological adaptability of Common Meadow-grass is also documented by the fact that it is one of the three flowering (higher) plants found in the Antarctic, where it was naturally introduced by man.

The genus *Poa* includes some 100 species. The name is derived from the Greek word *póa*, which originally meant any grass or herb.

Common or Smooth Meadow-grass is a stoloniferous, moderately tall herb. The leaf sheaths are rounded or slightly compressed, the ligules short and blunt. The leaf blades are rough on the edge and primary vein and ending in a point somewhat resembling the tip of a canoe (1). The inflorescence is a pyramidal panicle, sometimes quite long, with rough lateral branches.

The flowers (2) are in multiflowered spikelets and the base of the bloom is covered with long, shaggy hairs that sometimes look like the threads of spider-webs; the ripe grains (3) are often caught in the tangle of these hairs when they fall.

The flowering period is from late May to early August.

3
1

Wild Carrot
Daucus carota L.

Umbelliferae

The carrot is a food plant which was known to the Neolithic farmer, for early on man had noticed the tendency of the spindle-shaped root to thicken abnormally. During this process the phloem tissues of the vascular bundles become thin-walled and turn a deep orange colour. These tissues contain large quantities of sugar and the orange pigment carotene, which is changed into vitamin A in the body. The roots of cultivated carrots also contain vitamins C and B. Through the agency of man the simple, ordinary wild meadow carrot became an important vegetable valuable to his health. It is probable that the ancestor of the cultivated carrots was a hybrid of two original strains (subsp. *carota* and subsp. *maximus*), probably produced spontaneously in the wild. Cultivated carrots, of course, are very variable and the roots, which may be cylindrical with short or long tips, differ in taste, tenderness as well as sugar content.

In the centre of the white inflorescence (an umbel) there is often one or more purple flowers. Formerly it was believed that this central flower served as visual bait for insect pollinators; this was followed soon after by the contrary opinion that it repelled them. Lengthy experiments and observations confirmed neither of these hypotheses. The dark central flower neither attracts nor repels insects, either visually or chemically. It is only a kind of evolutionary peculiarity.

Wild Carrot is a biennial or plural-yearly herb. The pale turnip-like roots (1) as well as the roots of cultivated forms (2) contain faintly alkaline mineral substances, sugars and vitamins (carotene and hydrocarotene), which are important — in the raw state — to man's diet. The fruits (3) can also be used for medicinal purposes for they contain essential oils with anthelminthic (destructive to intestinal worms) properties. The foliage contains alkaloids. The leaves of Wild Carrot are pinnate and triangular-ovate in outline, the lower leaves are stalked, the upper leaves with small sheaths are stalkless, sessile.

The flowers (4), arranged in an umbel, are creamy-white (with the exception of the central flower) and either hermaphroditic or only male. When they are spent the drying umbel curls up into a ball (5). The fruit is an achene with ribs covered with long, hooked spines.

The flowering period is from May to August.

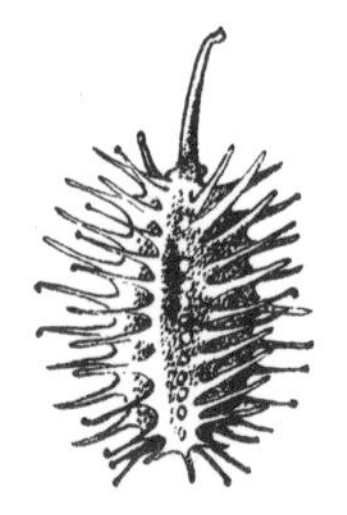

3

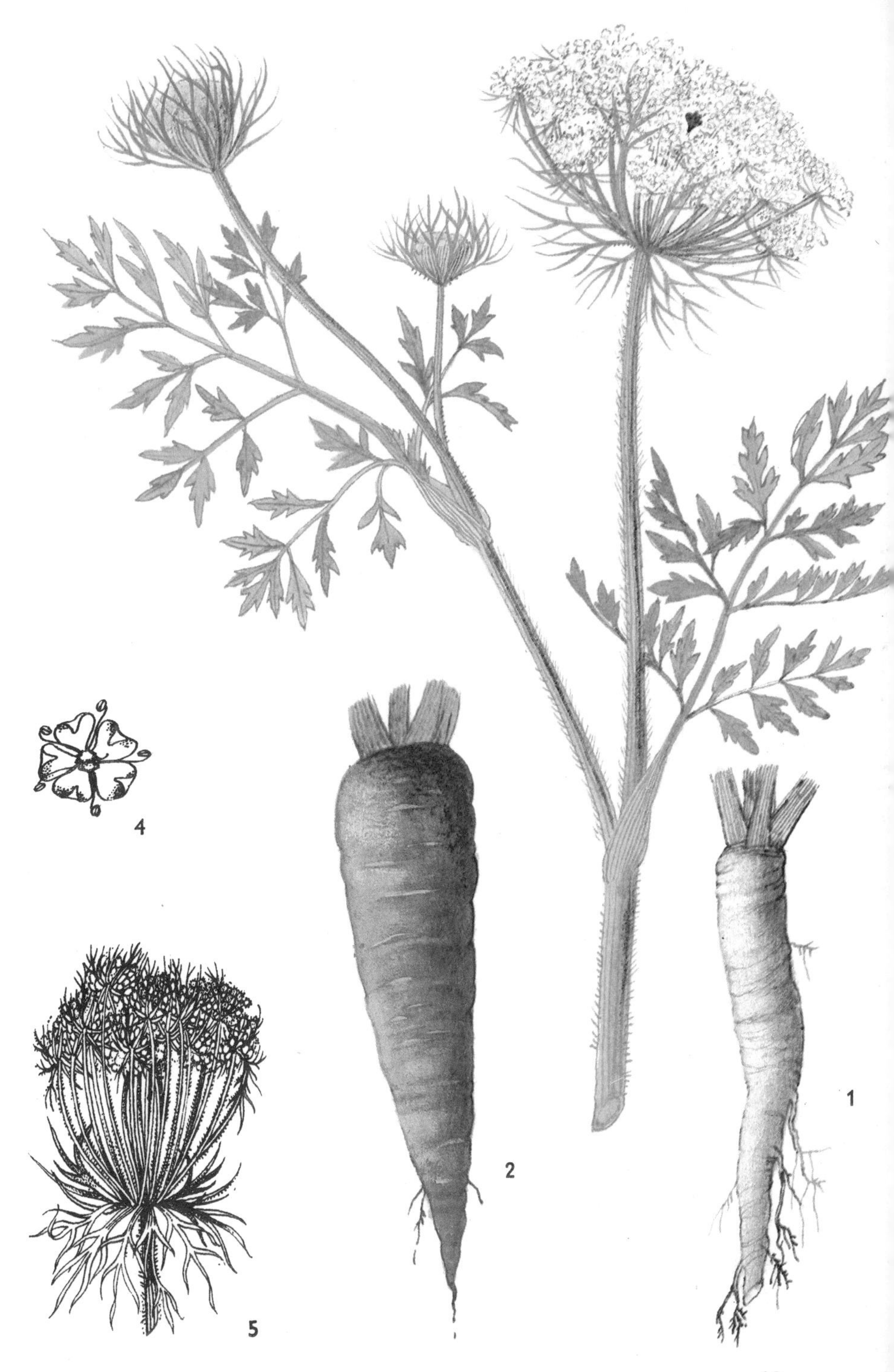
4
5
2
1

Spreading Bellflower
Campanula patula L.

Campanulaceae

From the economic viewpoint the best meadows are rather dry ones that can be cut for hay two or three times a year. They are usually found on rich, slightly damp, mineral soils in lowland districts and at lower mountain elevations. Of the many plants growing in such meadows the most colourful are Ox-eye Daisies, Geraniums — and — Spreading Bellflowers. These are distributed throughout the whole of Europe, excepting the extreme south and arctic regions. They also may be absent in isolated places, like the north German plateau, where if they do occur it is assumed they are not indigenous but were introduced.

Spreading Bellflower is a very variable species, the individual populations differing, for example, in the pubescence of the leaves or size of the flowers.

Bellflowers, the illustrated species included, are of interest biologically because of the maturation time of the male and female reproductive organs. They are examples of proterandry, that is having the stamens (or male organs) mature before the pistils (or female organs) in the same flower. In such plants self-pollination is thus practically ruled out.

Morphologically they are noteworthy for their fruits — capsules (1) that open by three pores; the pores expand in dry weather and slightly contract in damp conditions. In windy weather the seeds fall out through these pores.

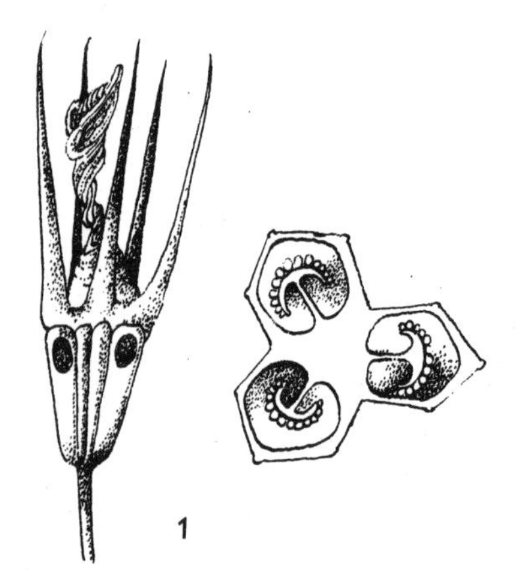

Spreading Bellflower is a biennial and often also a perennial herb, 15—50 cm high, with a basal rosette of oblong, crenate, short-stalked leaves; the leaves on the angled stem are sessile and diminish in size towards the top of the stem until they suddenly give way to involucral bracts. The small, typically bell-shaped flowers have long stalks and are arranged in loose, spreading panicles.

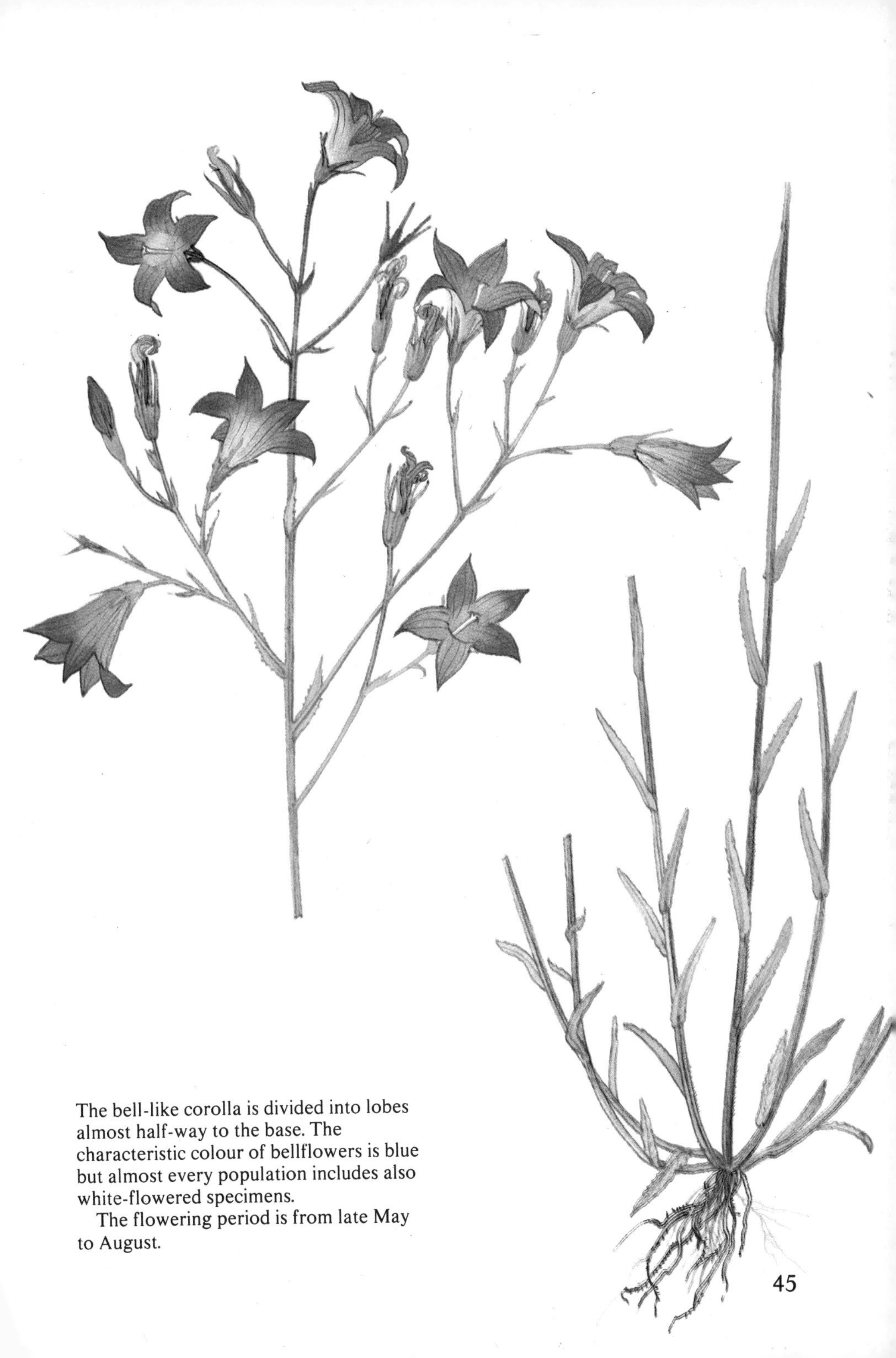

The bell-like corolla is divided into lobes almost half-way to the base. The characteristic colour of bellflowers is blue but almost every population includes also white-flowered specimens.

The flowering period is from late May to August.

Timothy
Phleum pratense L.

Gramineae

Timothy was formerly an ever-present natural component of valley and hillside meadows as well as open woodland clearings, waysides and hedgerows. It is found from lowland to high mountain elevations — in the Alps it has been encountered even at an altitude of 2,600 m — in continental Europe, northern Asia (excepting arctic regions) and North America.

Nowadays it is one of the most important of the grasses grown for forage and also to cover untilled tracts. For this reason it has been the subject of breeding and experimentation concerned not only with the production of seed but first and foremost with the uniform production of green matter and the hardiness and firmness of the turf. In the past years Holland alone, for example, grew five varieties of 'pastureland' Timothy, three meadow varieties and five transitional varieties. Like attention to breeding is given by the German Federal Republic, Denmark, Belgium and Great Britain (Welsh Plant Breeding Station, Aberystwyth). The so-called meadow types are early forms whilst the pastureland types are late forms on the evolutionary scale. The individual varieties also differ in appearance: meadow types make erect, compact tufts, pastureland types make more spreading tufts with some prostrate leaves. Important differences were determined also in the number of chromosomes.

Timothy is a plural-yearly grass with a moderately rough, distinctly furrowed leaf blade 5—8 mm wide. The ligule is 2—3 mm long and irregularly toothed. The leaf sheaths are rounded and sometimes bulbous at the base.

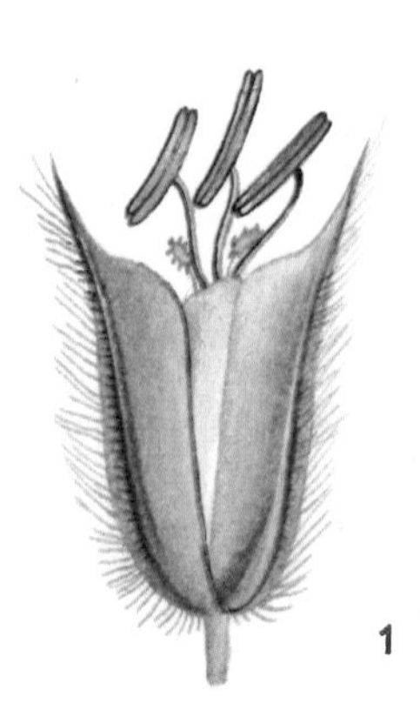

The culms are smooth and up to 100 cm high. The dense, cylindrical spike-like inflorescence is composed of a great many, densely clustered, one-flowered spikelets (1). Unlike the earlier-flowering Cat's-tail the spikelets cannot be stripped from the central stalk. The flowering period is from mid-June to mid-July.

The inflorescence of the similar Böhmer's Cat's-tail (*P. phleoides* (L.) Karsten) — which unlike Timothy grows on warm, sunny banks — forms lobes along the outer edge when bent over the finger (2) and this is thus an excellent means of identification.

2

Rough Hawksbeard
Crepis biennis L.

Compositae

Poor, slightly damp meadows as well as ditches and waysides (sometimes even rather dry embankments) from lowland to mountain districts is where you will encounter the yellow-flowered Rough Hawksbeard. It is not welcomed in the meadow by some farmers because its hay is tough and bitter, but it is valued by apiarists for it provides good food for bees. In early-mown meadows Rough Hawksbeard sometimes escapes notice because it normally flowers quite late and is thus cut before the flowers open. However it immediately puts out new shoots and forms numerous short-branched stems that are strikingly different from 'spring' plants. If it does bear flowers it produces ample seeds and sometimes becomes a weed of field crops. It is a typical European plant which has made its way together with field and meadow cultures to North America.

Plants of the genus *Crepis* have been the subject of intensive study by geneticists, who discovered not only variability in the number of chromosomes within a single species (*C. tectorum* L.), but also mutual sterility when south-European and Danish — in other words geographically distant — plants (*C. capillaris* (L.) Wallr.) were cross-bred. Short-lived hawksbeards are often common weeds and some scientists believe they are specialized products of the evolution of ancient perennial species.

Rough Hawksbeard is a moderately high herb, occasionally annual but usually biennial. The stems are stiff, grooved and branched at the top (spring plants). The basal leaves are longish-lanceolate narrowing into a stalk, the stem leaves are sessile, irregularly deeply toothed and lobed, with a large prominent terminal segment.

The small golden-yellow strap-shaped flowers (1) are clustered in a head (anthodium).

The fruit is a ribbed achene about 4 mm large with snow-white pappus.

The flowering period is from June to September.

Meadow Cranesbill
Geranium pratense L.

Geraniaceae

It is hard to imagine fresh, damp meadows without cranesbills. When the flowers have faded they are a striking feature with their long-beaked, schizocarpic fruits often likened to cranes' bills, hence the plant's common name. The awns of the beaked fruits, each attached to a one-seeded fruit segment, separate from the prolonged receptacle and coil sharply. They are relatively hygroscopic and straighten again in damp weather. This property is retained for some time and thus the awns are a kind of 'natural' hygroscope.

Plants of the geranium family are found chiefly in temperate regions. Of the approximately 300 species of the genus *Geranium* the following three are the ones most often encountered in meadows: Meadow Cranesbill in rather dry, fresh meadows and rich soil — this is a Eurasian species whose numbers are declining in the north; the Marsh Cranesbill (*G. palustre* L.) by stream margins, in damp meadows and in shoreline thickets in the suboceanic parts of Europe; the Wood Cranesbill (*G. sylvaticum* L.) in fertile mountain meadows and in the neighbourhood of springs. The separate species have very close ties, as testified to also by the chemico-taxonomic studies of the substances they contain, chiefly tannins and sugars. There are greater differences in the concentration of these substances within a single plant during the growing season than there are between the individual species.

Cranesbills are annual or perennial herbs. The generic name *Geranium* is derived from the Greek word *géranos*, meaning crane — perhaps also because of the beaked fruit.

The Meadow Cranesbill (1a, 1b) is a perennial herb with short rhizome, erect, hairy stem and deep five- to seven-lobed leaves (the upper leaves only three- to five-lobed). The flowers are pale blue-violet, very occasionally pink or white, and are produced from late June to September.

The very similar Wood Cranesbill (2a, 2b) is also a perennial herb with five- to seven-lobed hairy leaves; the lobes are

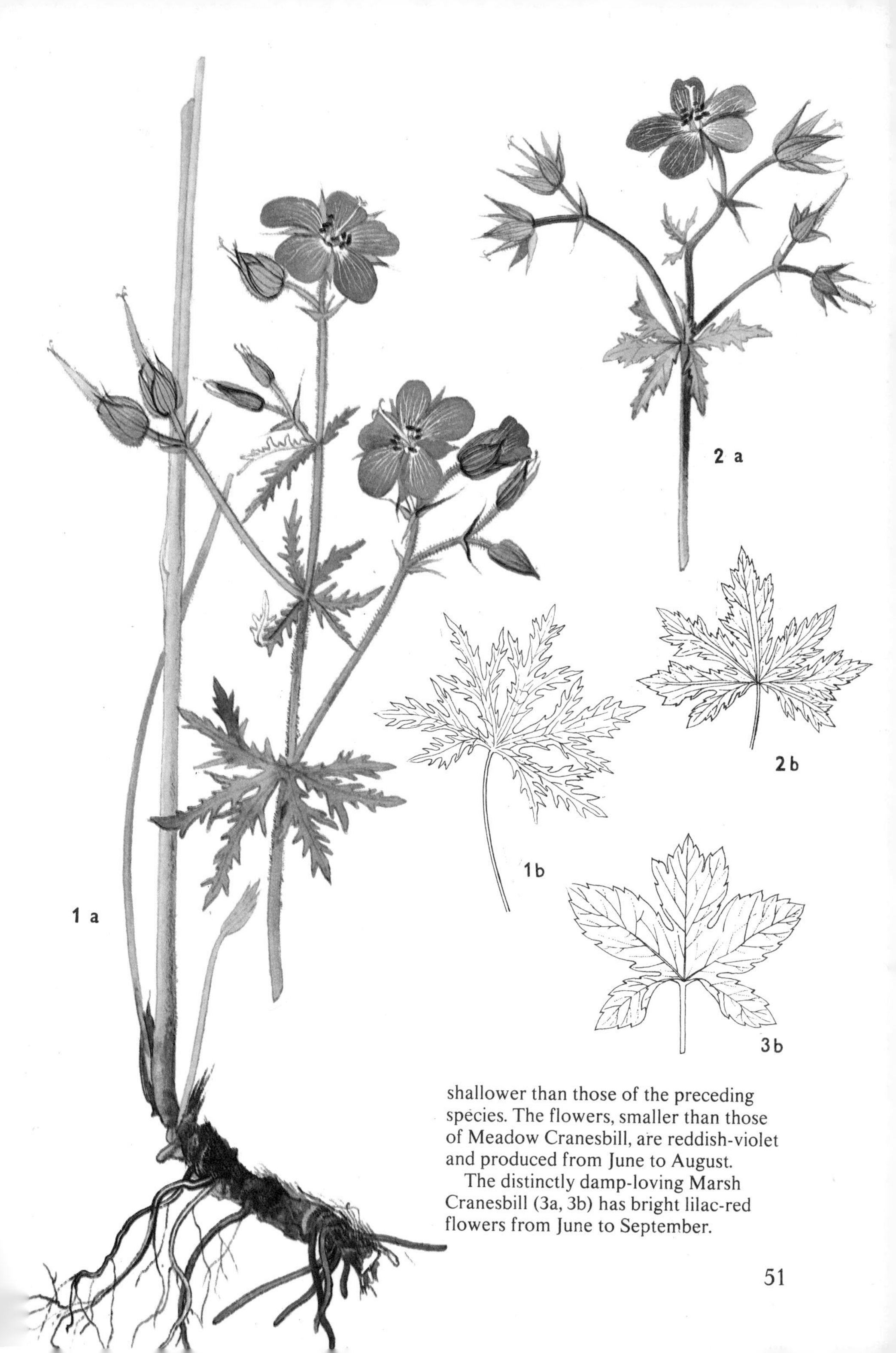

shallower than those of the preceding species. The flowers, smaller than those of Meadow Cranesbill, are reddish-violet and produced from June to August.

The distinctly damp-loving Marsh Cranesbill (3a, 3b) has bright lilac-red flowers from June to September.

Tall Oat-grass

Gramineae

Arrhenatherum elatius (L.) J. & K. Presl

Oat-grass was always viewed as a kind of oat. In fact it has a similar inflorescence and many other common features in the flower — for example awns and styles with feathery stigmas. The long awns protruding from the flowers are also the characteristic that gives the genus *Arrhenatherum* its name, from the Greek words *arrén,* meaning male, and *athér,* meaning awn.

The tall culms of Oat-grass cannot be overlooked in any meadow. Oat-grass has even become the symbol of the meadow; entire communities of good, economically exploited meadows on fresh as well as drier soils are classed as 'Oat-grass meadows' — in every instance they are fertilized, often-mown meadows, dependent on man's activity.

Oat-grass would grow in the wild, of course, without man. It was a characteristic species of natural, fresh, mesophytic meadows where it grew together with Meadow Cranesbill, Hogweed, Spreading Bellflower, Wild Carrot, Hawksbeard and other species. It is also found in woodland clearings, grassy banks, and hedgerows from lowland to foothill districts and in mountain valleys up to elevations of 1,300 m. Because it does not tolerate trampling by livestock it is usually absent in pastureland. It is suitable for adding to clover-grass mixtures but is not palatable to cattle by itself for it is faintly bitter.

It is distributed in practically all of Europe, north Africa, the Canary Islands and central and western Asia; it was also introduced to North America and Australia.

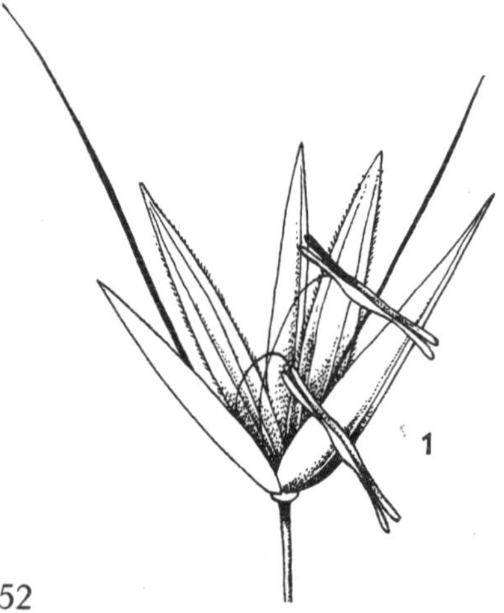

Tall Oat-grass is one of the tallest meadow grasses, reaching a height of 60—180 cm. It has a deep root system, which helps it stand up well to dry conditions or temporary drying up. It is a loosely-tufted perennial grass with erect, shining culms. The leaf sheaths are rough, the ligules short and blunt. The inflorescence is a panicle, long and narrow at first, later spreading. Its branches are short and rough. The spikelets are usually two-flowered; the lower floret is male with a long awn and

two stamens that often 'hang out' from the flower (1); the upper floret is hermaphroditic and awnless. The bottom lemma has a jointed awn up to 10 mm long.

The flowering period is from June to the beginning of August.

Ox-eye Daisy
Chrysanthemum leucanthemum L.

Compositae

When meadows are abloom with white Ox-eye Daisies it's a sign that summer is just around the corner. In fact the outer ray florets are not white but colourless. The white is the result of the reflection of light by the air contained in the intercellular spaces of the ray florets. The central disc florets are golden-yellow and the number of outer ray florets is extremely variable. As early as the beginning of the 20th century one English botanist observed that there may be anywhere between 11 and 35, but the usual number was 22. Further investigations revealed that the differences in the number of ray florets were linked with the time the flowers were picked and were also determined by the local strain.

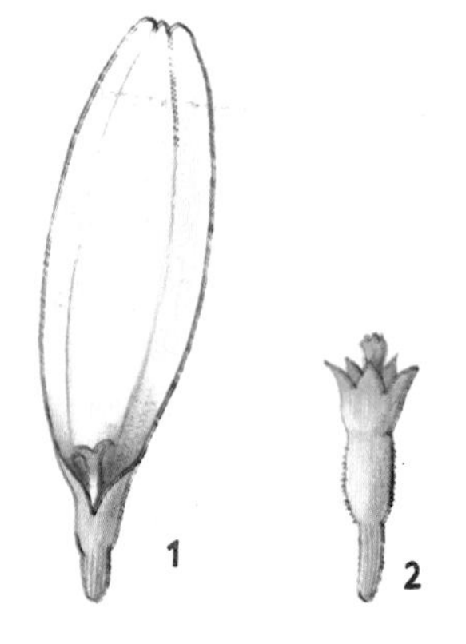

Ox-eye Daisy, which grows in freshly-damp as well as dry meadows, is one of 350 species of chrysanthemums although it is also classified under the separate genus *Leucanthemum* under the name *Leucanthemum vulgare* Lam. It is a perennial herb up to 80 cm high, hairy as well as glabrescent, with long-stalked basal leaves that are obovate in outline, coarsely toothed on the margin and sessile. The stem leaves are longish-ovate in outline and coarsely serrate on the margin.

The inflorescence (3) of Ox-eye Daisies, composed of ray (1) and disc (2) florets, is usually 3—6 cm across, but the inflorescence of garden varieties is sometimes as much as 15 cm in diameter. Some garden varieties have double flowers, that means they have more than the normal number of white ray flowers. The fruit is an achene (cypsela).

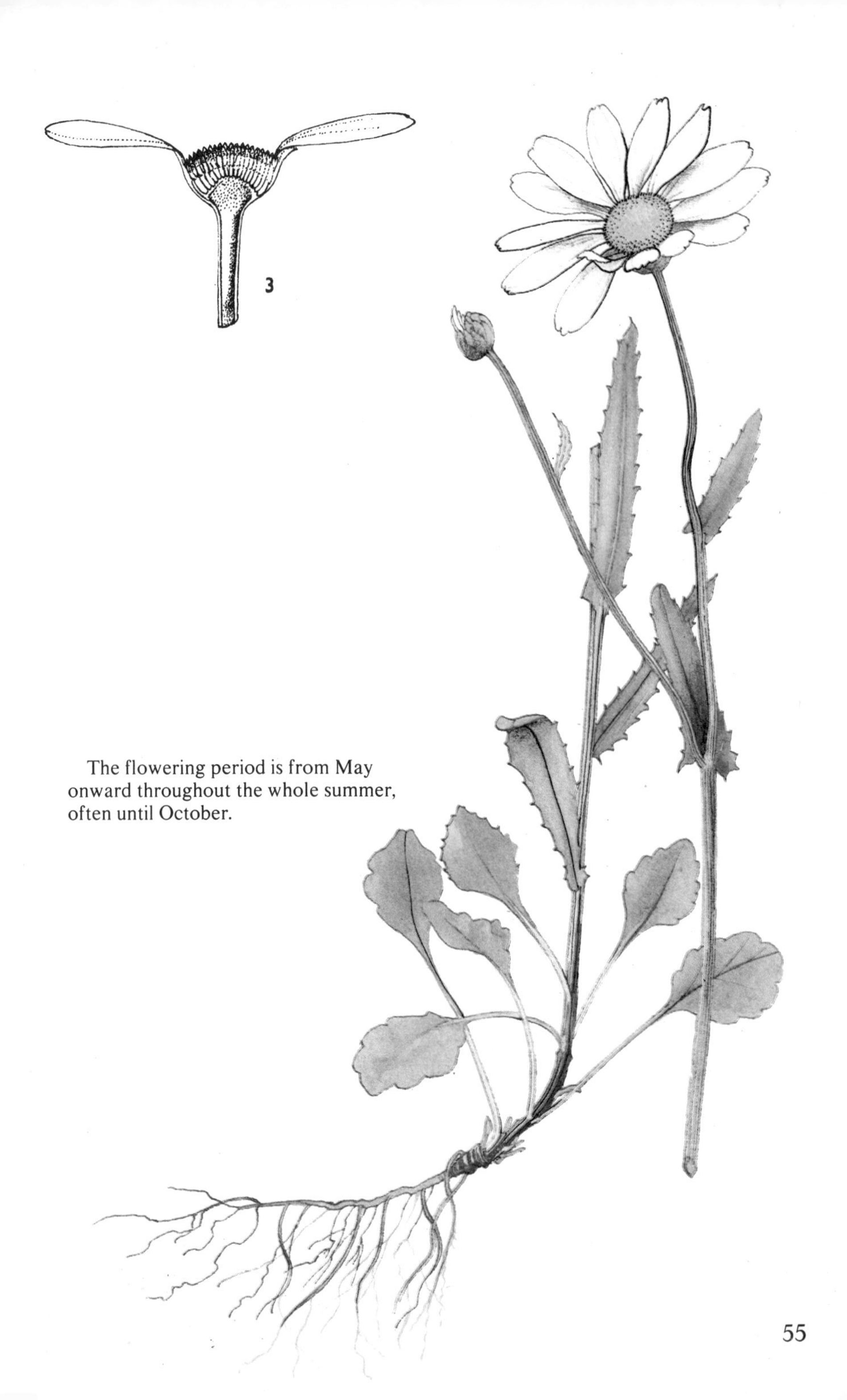

The flowering period is from May onward throughout the whole summer, often until October.

Hedge Bedstraw

Galium mollugo L.

Rubiaceae

Hedge Bedstraw is a common plant of fresh meadows. Its long decumbent stems, often more than 1 m in length, made work difficult for mowers in days when the scythe and plough were the main farming implements. These long, decumbent stems are capable of rapidly righting themselves when laid flat by a downpour or hail, a property usually attributed only to grasses. This phenomenon is provoked by gravity and is known as negative geotropism; unlike grasses, however, the meristematic tissues are not localized in joints (nodes).

The white-flowered Hedge Bedstraw often grows together with the yellow-flowered Lady's Bedstraw (*Galium verum* L.). South-east European races of the two do not interbreed. In central and northern Europe, on the other hand, the two occur as genetically different strains that readily interbreed and in places where Hedge and Lady's Bedstraw both grow one may encounter fertile hybrids of the two.

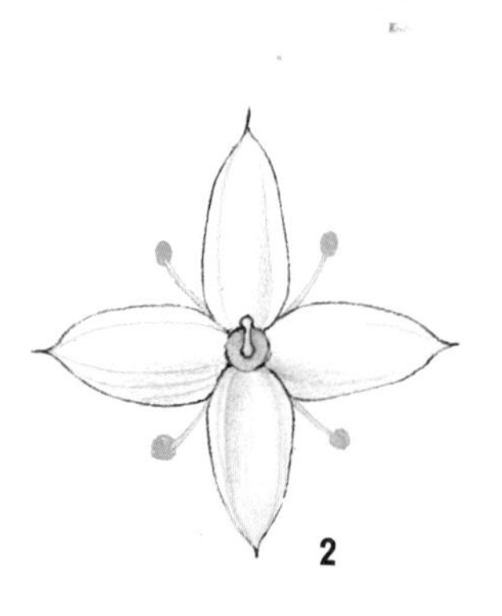

Some 300 species of bedstraw are distributed throughout the world excepting Australia. *G. mollugo* and *G. verum* are perennial herbs up to 1 m high. Hedge Bedstraw (1) has a branched, creeping rhizome and rounded four-angled, decumbent stem with whorls of four to eight linear leaves. The four-lobed flowers are usually creamy-white and 2—4 mm in diameter (2).

The yellow-flowered Lady's Bedstraw (3) has erect stems with whorls of up to 12 narrowly-linear leaves that are felted on the underside.

Bedstraws flower from June, Lady's Bedstraw from July, to September.

1
3

Lady's Smock
Cardamine pratensis L.

Cruciferae

Every meadow that is at least slightly damp glows in spring with the white, rosy-violet and pinkish flowers of thousands of Lady's Smocks. It is a very adaptable plant and its characteristic basal rosette may be encountered even in warmer and drier places. This adaptability is also testified to by its distribution. It is one of the few circumpolar species and is found throughout the Eurasian continent excepting southern Spain, Italy and the Balkans, the southern limits of its range extend eastward through the fountainheads of the large Siberian rivers as far as Sakhalin. In North America it is a common plant of meadows, particularly in Canada and the northern United States. It occurs in isolation on the coast of Greenland and in Spitzbergen. Not many of the world's plants have such a widespread distribution.

This is probably connected with the practically cosmopolitan distribution of the entire crucifer family which tolerates extreme environmental conditions. *Braya purpurescens* Bunge, for example, grows in Grinell Land (in the Arctic Circle) at 83° 24′ latitude North and *Ermania koelzii* O. E. Schulz was found in Kashmir at 6,300 m above sea level. Members of the crucifer family are often the last vascular plants in high mountains and many are man's companions, as weeds, on all continents.

Lady's Smock is a perennial herb with short rhizome and erect, little-branched (or unbranched) stem. The basal leaves, often lasting the winter, are arranged in a striking rosette sometimes completely pressed to the ground (1). These leaves are odd-pinnate with ovate-oval leaflets. The stem leaves are short-stalked and placed at a sharp angle.

The flowers (2) are usually pink or white with darker violet veins; the anthers are generally pale yellow. The fruit is a siliqua with straight stalk.

The flowering period is from April to June.

2

Creeping Jenny
Lysimachia nummularia L.

Primulaceae

Creeping Jenny grows in damp meadows, by the waterside and in woodland ditches from lowland to foothill districts. It has a continuous distribution embracing all of Europe — as far as 62° latitude North — from England to the USSR excepting the southern Mediterranean region. It has also been introduced to the Atlantic regions of North America and to Japan.

Research pertaining to the medicinal properties of this plant has shown that it aids in suppressing the flu virus. In old herbals as well as in more recent pharmacopoeias, however, the infusion from the top parts of the plant is recommended for diarrhoea and as a tonic. External application of the macerated fresh plants helps heal wounds, alleviate pain in rheumatism (inflammation) of the joints and has a beneficial cosmetic effect on the skin (it makes it supple).

Common in moss-moor meadows, by inlets, streams, ponds and rivers is the related, but upright, Yellow Loosestrife (*Lysimachia vulgaris* L.) and round springs in foothill districts Dotted Loosestrife (*Lysimachia punctata* L.) with lemon-yellow flowers.

Some loosestrifes (including Creeping Jenny) are grown in the garden: the variety 'Aurea' with golden-yellow flowers has proved to be a good carpeting plant for damp sunny places.

Creeping Jenny (1) is a creeping, perennial herb, usually pressed close to the ground, with a four-angled, little-branched stem that roots at the nodes. The leaves are opposite, rounded, coin-like with faintly heart-shaped base. The fruit is a capsule but the plant spreads and multiplies more by vegetative means than by seeds. The top parts contain saponins, tannins, silicic acid

and phytoncids; research on further constituents is continuing.

Yellow Loosestrife (2) is a robust perennial herb up to 1 m high with loosely branched stem and leaves either opposite or in whorls of three or four leaves. The flowers are arranged in terminal panicles and also grow from the axils of the upper leaves.

Creeping Jenny flowers from May to July, Yellow Loosestrife about a month later.

Cow Parsley

Anthriscus sylvestris (L.) Hoffm.

Umbelliferae

Even such an ordinary, almost weedy herb as Cow Parsley can be beautiful! Perhaps not as a single plant but in spreading masses, as in Bavaria, it is truly a sight to behold.

Such meadow communities are not the only place where Cow Parsley grows. In the modern landscape, so greatly changed by the inroads of civilization, it has found a place for itself in ruderal spreads of Goutweed and Burdock. In recent years it is most abundant in ditches alongside highways. They are ecologically suitable (damp) and economically unexploited (the grass growing there is no longer cut because of its high lead content) and, therefore, afford excellent conditions for its spread. It is even making its way into towns via such untended roadsides. It also grows in damp, nitrogen-rich woodlands, e.g. alder groves.

Not too long ago the related *Anthriscus cerefolium* subsp. *sativum* was grown and widely used throughout Europe as a flavouring for soups and sauces.

Cow Parsley is a Eurosiberian species found in central and northern Europe and extending through Siberia far to the east. It is also naturalized in North America where it was introduced with shipments of seeds.

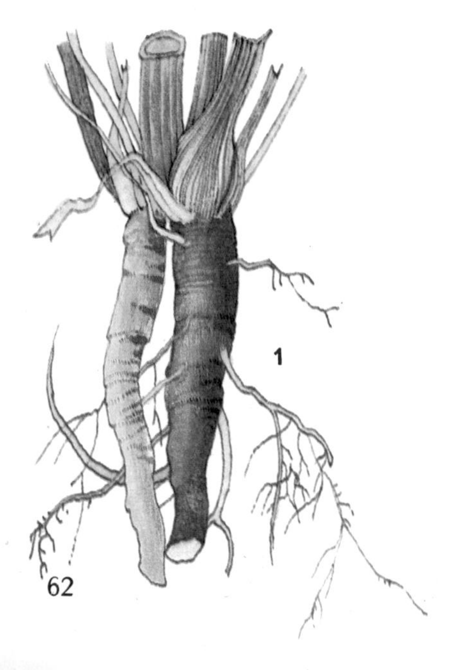

Cow Parsley is a biennial or plural-yearly herb with stout rhizome (1) that overwinters along with the basal leaf rosettes.

The stems are hollow. The compound, two- to three-pinnate leaves are triangular-ovate in outline. The flowers are white, tinged with green or yellow, and shallowly notched. The marginal flowers in the umbel are faintly radiate, i.e. the outer petal is strap-shaped (2). The fruit is a conical, compressed, shortly-beaked achene (3).

The flowering period is from May to July.

2
3

Common Sorrel
Rumex acetosa L.

Polygonaceae

Common Sorrel is a common, widespread herb of damp, natural as well as artificial meadows; however, it is very adaptable and is also found in drier places — pasturelands, hedgerows as well as on railway embankments from lowland to mountain elevations (approximately 2,000 m in the Alps). Its natural area of distribution is large and circumpolar. It is found in practically all of Europe and Asia, on the east and west coast of North America, in Canada and in Alaska.

The plant's sour taste, its acidity, is caused by the high concentration of oxalic acid, which may be poisonous in large quantities. In some countries (e.g. France) it is used to make salad and it is there that cases of poisoning are recorded. A similar effect may be provoked by eating a large amount of rhubarb — which belongs to the same family — whose stalks are used to make an excellent filling for pies. Sorrel leaves, as well as rhubarb, should also be avoided by people whose bodies tend to form gravel (small oxalate concretions) in their kidneys.

The concentration of oxalic acid in the tissues of Sorrel is so great that sap pressed from the plant makes litmus paper rapidly turn red. When boiled briefly the green leaves immediately turn golden-yellow or brown: the oxalic acid in the dead cells changes the green colouring matter into the pigment pheophytin.

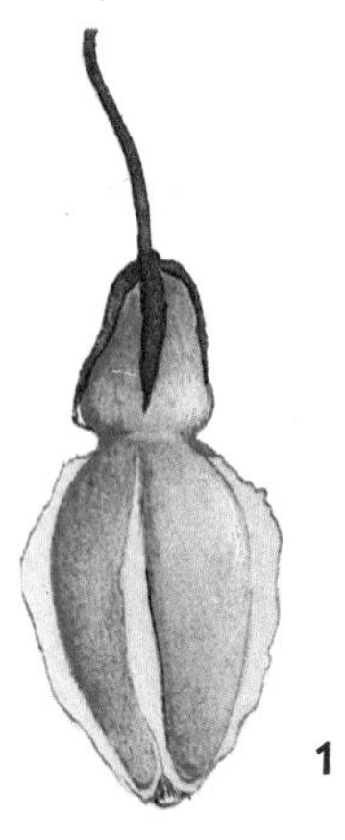
1

Common Sorrel is a perennial herb approximately 50—60 cm high with longish arrow-shaped leaves; the basal leaves are long-stalked, the stem leaves sessile.

The flowers are clustered in a panicle-like inflorescence sometimes up to 25 cm high. The structure of the flowers is unusual: the perianth is composed of 6 segments arranged in two rings; the inner segments become conspicuously enlarged and membranous and remain on the stalk in the fruit. These 'wing-like' segments aid in the dispersal of the fruit either by wind or water and their incipient stage is visible already in the bud (1). They are also an important

means of identification: they are usually furnished with appendages whose arrangement is characteristic for the various species of the genus *Rumex.* The fruit is a three-sided achene (2).

The flowering period is from May to July.

Meadow Buttercup
Ranunculus acris L.

Ranunculaceae

Meadow Buttercup is a typical plant of moderately damp meadows, particularly in alluvium deposited by rivers and streams. Characteristic of such meadows is the permanent presence of variegated herbs amidst the grass — like ox-eye daisies, clovers, sorrels, burnets or buttercups. Most like these natural meadows are the nutrient-rich artificial meadows of central Europe. At a certain time of the year the yellow of buttercups predominates over all else in these meadows — partly because their flowers remain open a relatively long time (a single blossom of the Meadow Buttercup as long as seven days).

Meadow Buttercup is an example of a plant distinguished by marked adaptability. In the middle of the present century scientists in England studied tracts with Creeping Buttercup (*R. repens* L.), Bulbous Buttercup (*R. bulbosus* L.) and Meadow Buttercup. It was discovered that the presence and quantity of these three species depended on the soil drainage: Creeping Buttercup was most common in depressions and inundated furrows; Bulbous Buttercup, on the other hand, predominated in elevated and well-drained places; whilst Meadow Buttercup exhibited no special preference and was found in all types of places. The same is true of germination. Bulbous Buttercup germinates in rather dry soil; Creeping Buttercup in moderately wet and flooded soil and Meadow Buttercup somewhere inbetween. This should be kept in mind when looking for them in the meadow.

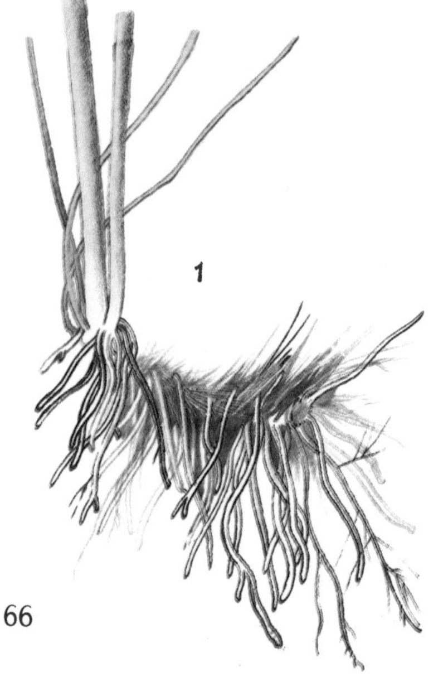

Meadow Buttercup is distributed throughout practically all of Europe and north-western Asia, also in several places in north Africa and on the coast of Greenland. However, it is a very variable plant and readily interbreeds with other buttercups. It may be anywhere from 10 to 100 cm high and covered with appressed hairs as well as glabrous. Concealed in the ground is a short, stout rhizome (1) from which rises an erect, branched stem. The basal leaves are long-stalked and palmatifid with five to seven tripartite lobes. The stem leaves are sessile and similar to the basal leaves but with narrower to linear segments. The fruit is an achene with a short beak (2).

Meadow Buttercup is considered to be a poisonous plant.

The flowering period is from May to September.

Ribwort
Plantago lanceolata L.

Plantaginaceae

Ribwort is a meadow herb that figures in ancient tales and legends. Its leaves have been used as a medicinal drug since time immemorial, though pharmacological study of it still continues. In folk medicine it was used primarily in the form of compresses made from the crushed leaves and applied to suppurating and stubborn wounds, as an expectorant (easing the bringing up of mucus from the respiratory tract) and as an infusion to regulate bowel movements.

Nowadays Ribwort is a cosmopolitan plant. Originally it grew only in Europe and northern and central Asia but it was introduced to other continents — to Australia, north Africa and North America; perhaps it was brought there by settlers who took it with them as a valuable medicinal plant.

The seeds of Ribwort (as many as 5,300 have been counted on a single plant!) are used by bird-keepers to feed birds. The seed coat swells in damp conditions and sticks to the feet of passers-by, thereby aiding the dispersal of the seeds. This is the commonest means of the plant's distribution, particularly in view of the fact that the seeds remain fertile for as long as 11 years. If Ribwort thus finds its way into field crops it may become a troublesome weed. The fresh leaves, when eaten by livestock, may cause diarrhoea.

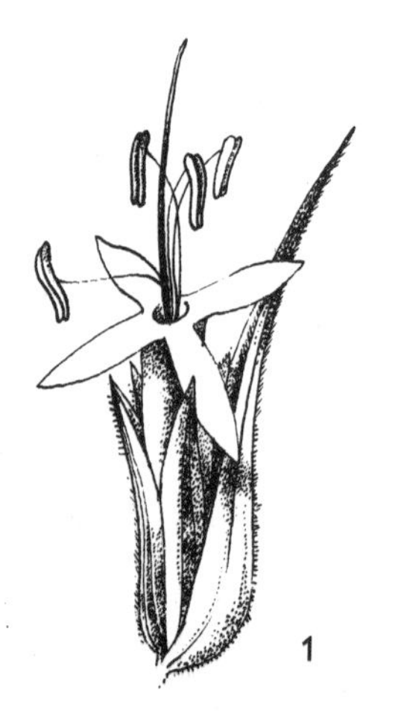

The scientific name *Plantago* is supposedly derived from the Latin word *planta,* meaning both plant sprout and (primarily) sole of the foot, which is what the leaves appeared to resemble to our European ancestors. Of interest, however, is the fact that the same name was arrived at independently by the American Indians. They called plantains introduced to North America White Man's Footprint, not only because of the leaf shape but also because plantains stand up well to trampling and often grow on field paths.

Ribwort is a perennial herb with lanceolate leaves arranged in a basal rosette and scape 5—50 cm high bearing tiny 4-merous flowers (1). Besides other substances the leaves contain glycosides, tannins, and the like.

Ribwort flowers from May to September.

Tufted Vetch
Vicia cracca L.

Fabaceae

Tufted Vetch is one of the extremely variable species of the genus *Vicia.* This is doubtless due to its great ecological amplitude (it grows in damp moss-moor meadows as well as in grain crops and on dry rocks) and its geographical distribution, extending from Iceland through all of western and central Europe eastward to central Asia and the Far East, to Japan and middle China. A complex species, it embraces numerous minor taxons: subspecies, varieties and forms. Variability is a trait exhibited by most of its morphological characters; the plant's size, manner of branching, length of the flower stalk, shape of the inflorescence, number of flowers in the inflorescence, colour of the flowers, and size as well as number of seeds in the pod. Most of the various described forms, however, are determined by their habitat and if grown in cultivation the striking differences disappear or are not constant.

Vetches include a great many species that are of economic importance, i.e. grown for fodder. The annual Common Vetch (*V. sativa* L.) is popular with apiarists as food for bees; the cup-shaped nectaries, however, are not located in the flowers but on the leaves!

Tufted Vetch (1), one of the approximately 120 species of *Vicia,* is a perennial herb with creeping rhizome and stiff, angled, prostrate or climbing stems up to 1.5 m long. The leaves are even-pinnate and the terminal (odd) leaflet is modified into spirally-coiled tendrils.

The flowers are arranged in a one-sided racemose inflorescence (often bearing as many as 30 flowers) that becomes arched. The fruit is a pod 2—3 cm long. The flowering period is from June to August.

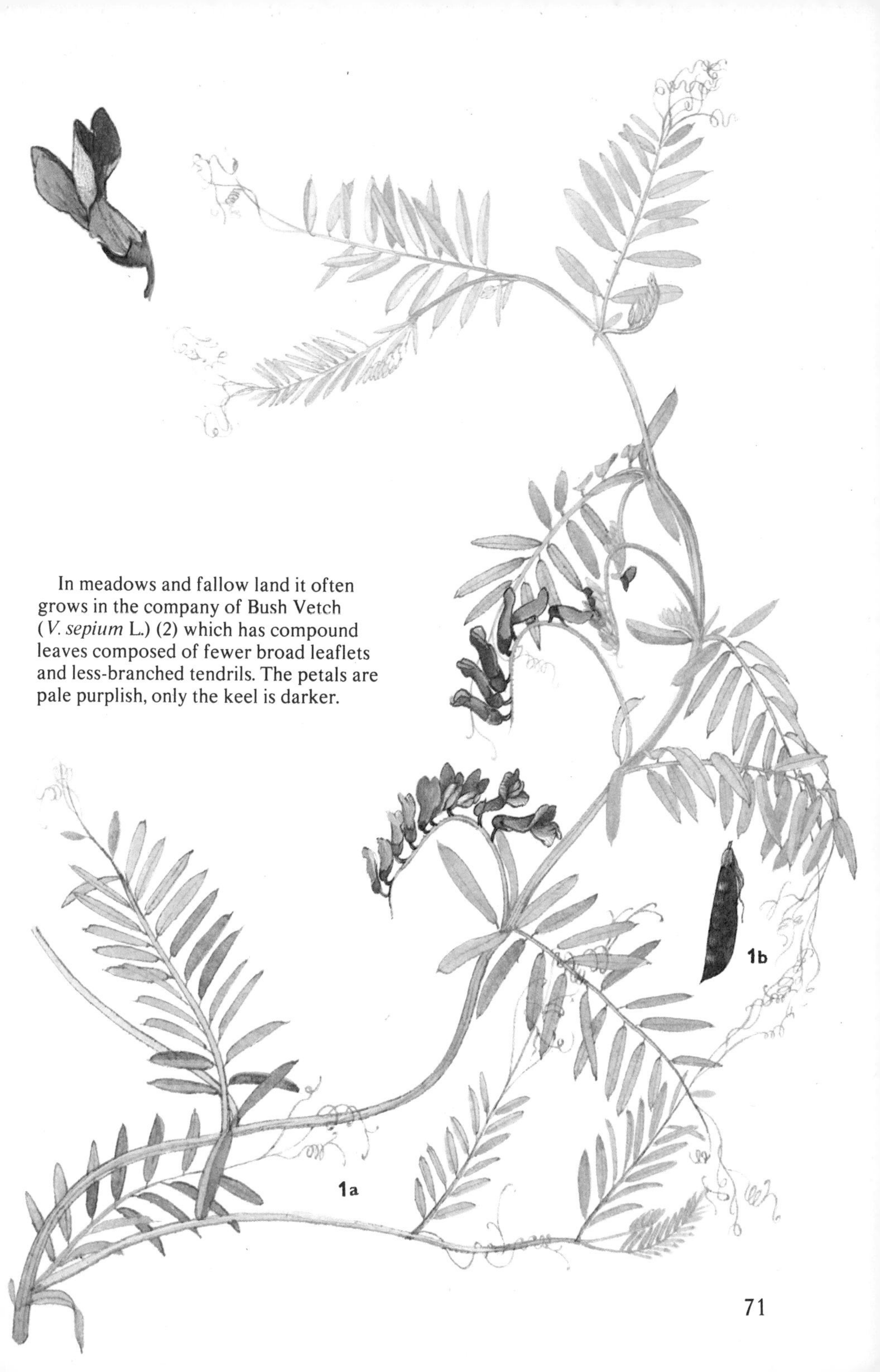

In meadows and fallow land it often grows in the company of Bush Vetch (*V. sepium* L.) (2) which has compound leaves composed of fewer broad leaflets and less-branched tendrils. The petals are pale purplish, only the keel is darker.

Hogweed

Umbelliferae

Heracleum sphondylium L.

Hogweeds are some of the largest herbs of the temperate regions. The illustrated *Heracleum sphondylium,* native to Europe, is not the biggest — only occasionally does it reach a height of 150 cm. It is an extremely variable herb, particularly in the shape of the leaves, which has led to the description of many intraspecific taxons. Irregularity, even monstrosity, is a trait exhibited also by the structure of the flowers; there have been instances of flowers with subtending bracts shaped like normal but miniature compound leaves.

The present day and age is witnessing the expansion of another hogweed throughout practically all of Europe — *Heracleum mantegazzianum* Somm. & Lev. — a herb up to 3 m high with leaves more than 1 m long and umbels $^1/_2$ m across. Because it, too, grows in damp places — often in ditches by roads and highways, along which the seeds spread most frequently — it is often believed to be an 'oversize' *H. sphondylium.*

On record are cases of serious skin burns 24 hours after coming in contact with the fresh stems of *H. mantegazzianum.* These skin allergies are called photodermatoses.

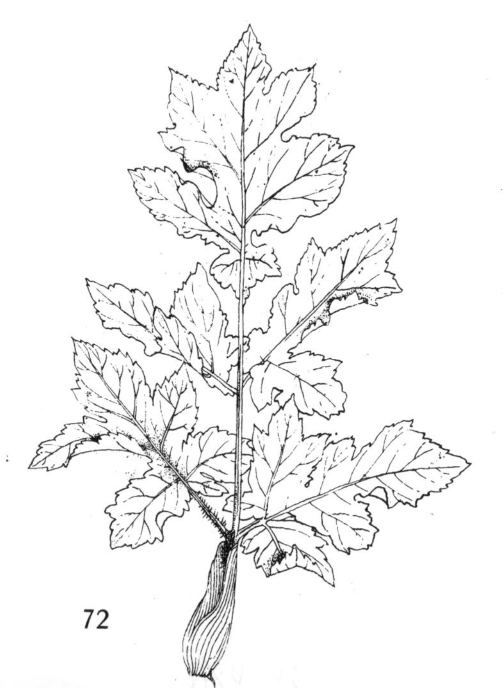

Hogweed is a tall, robust perennial herb and it is this that probably earned it its scientific name *Heracleum* (after Heracles, or Hercules, of Greek mythology).

The underground rhizome bears erect, hollow stems that are angled and grooved, covered with bristly hairs and branched at the top. The leaves are compound and of diverse shape.

The umbels at the tips of the branches are large, umbrella-like and generally composed of hermaphroditic flowers. The lateral umbels are usually smaller and composed only of male flowers. The flowers are white, but may be tinted cream, green or pink. The petals of the marginal flowers are conspicuously radiate (they are longer and more deeply notched than those of the other flowers) (1).

The flowering period is from June to August.

1

Yorkshire Fog
Holcus lanatus L.

Gramineae

Yorkshire Fog is a soft, silky grass that stands out, even in a mass of many other species of grass, because its softness is readily apparent to the eye.

Grasses of the genus *Holcus* have caused naturalists many a headache, however. In general there are two species in Europe — Yorkshire Fog and Creeping Soft-grass (*H. mollis* L.) — and it is very difficult to distinguish between the two. Yorkshire Fog is a uniform species with a continuous distribution throughout all of central and western Europe, excepting certain isolated localities in northern Scandinavia. It is also found in the north-western (Mediterranean) parts of north Africa. Creeping Soft-grass has a slightly smaller range but occurs in four races with differing number of chromosomes. In Great Britain, for example, Creeping Soft-grass occurs as a sterile race that perpetuates itself only by vegetative means. On the basis of extensive analyses the English botanist K. Joines arrived at the conclusion that Britain's Creeping Soft-grass is a very complex hybrid with a parentage that also includes Yorkshire Fog.

Yorkshire Fog is a typical meadow plant which grows in damp, lime-deficient and partly marshy meadows. Creeping Soft-grass, on the other hand, is a plant of forest margins, fallow land and fields. It is a component of numerous weed communities on arable land at median elevations (hilly country) from 500—1,000 m above sea level, particularly in acidic mountain soils.

The genus *Holcus* is not large, embracing barely 10 species. Yorkshire Fog is a densely-tufted, greyish-green grass 30—80 cm high. The culms are erect and softly hairy at the nodes. The leaf sheaths and blades are also hairy. The ligule is about 2 mm long and laciniate. The inflorescence is a branched panicle, the branches are likewise covered with hairs. The spikelets are two- to three-flowered and reddish (particularly before they open). The male flowers are furnished with a minute hooked awn that does not protrude much from the spikelet (1).

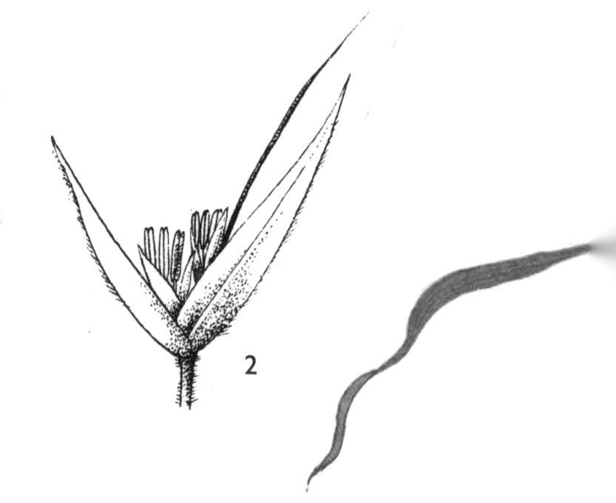

The very similar Creeping Soft-grass has a creeping, stoloniferous rhizome so that it does not form tufts. The awn in the male flowers is very long and protrudes from the flower (2).

The flowering period for both is from June to September.

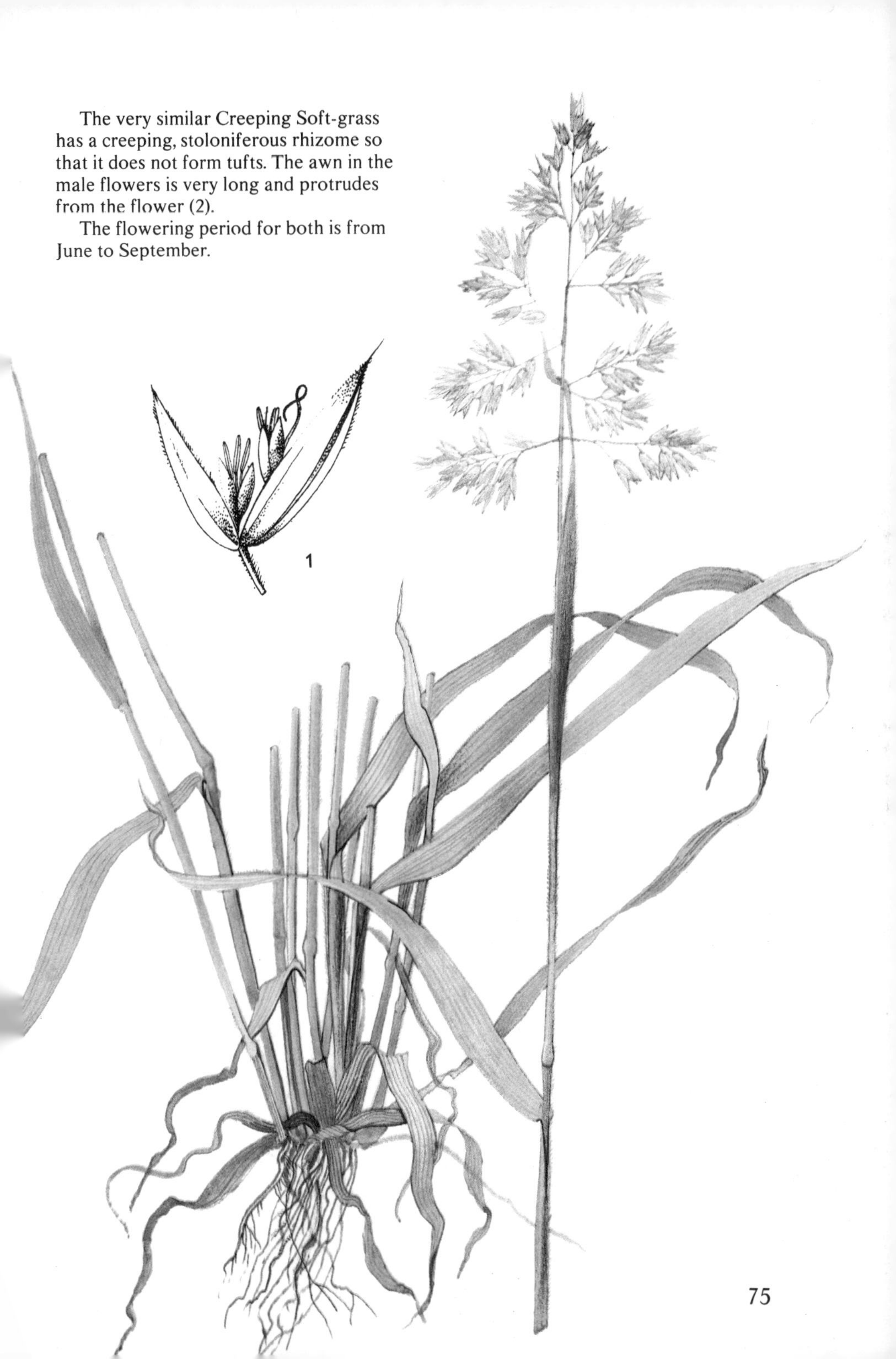

Wood Goldilocks
Ranunculus auricomus L.

Ranunculaceae

Whereas Meadow Buttercup is the yellow dominating element of moderately damp artificial meadows in spring, in damp to definitely wet meadows its place is taken by Wood Goldilocks, which also grows in deciduous woods and many parks where wet meadows alternate with groves and trees. It flowers in early spring (as early as April) and is a permanent component of the herb layer until the frost.

The flowers of Wood Goldilocks, like those of all buttercups, have a great many stamens. If the plants were carefully dissected, however, and the parts placed side by side, a continuous morphological scale would be seen ranging from the standard stamen with anther to normally developed, coloured petal. Some scientists deduce from this that the petals developed from the male reproductive organs during the course of evolution. In the continuous scale from stamen to petal the anthers (pollen sacs) become smaller, lose their fertility and the transitional organs begin to show signs of developing nectaries. This property, i.e. the modification of stamens into forms resembling petals, was noticed also by nurserymen and it is amongst plants of the buttercup family that they succeeded in obtaining, through selection, a great many double forms.

Wood Goldilocks is distributed throughout practically all of Europe (excepting the south and Alps), extending as far as Siberia. It is a perennial herb 15—50 cm high with short rhizome and erect, little-branched stem. The basal leaves are conspicuously different from the stem leaves. The leaf-shapes on a single plant may range from rounded kidney-shaped, shallowly toothed and long-stalked at the bottom, through deeply three- to five-partite, to sessile, palmatisect stem leaves with nearly entire, linear segments (1).

The flowers (2) on rounded stalks are golden-yellow (hence also the Latin name *auricomus*, meaning golden-haired, or goldilocks). The fruits (achenes) have a short, hooked beak.

The flowering period is in April and May.

1

Meadow Foxtail
Alopecurus pratensis L.

Gramineae

Foxtail is one of the most important of forage grasses. Extensive meadow communities with foxtail prevailing are common chiefly along the lower and middle reaches of rivers in lowland and hilly country. They are generally found on loamy-sandy soils enriched with nutrients by spring floods. The damming of water courses in the present century caused a marked decline in the number of natural foxtail meadows. Foxtail, however, has become an important cultivated meadow grass often included in seed mixtures for such meadows. It is relatively sensitive to nitrogen fertilizers and itself is an important source of nitrogenous compounds (plant proteins). During the growth period marked differences were observed in the growth dynamics of foxtail grasslands — unlike other grasses it is tallest in spring. This goes hand in hand with its general, overall early development: the flowers also appear early thereby providing it with ample opportunities to seed itself.

In recent decades foxtails are occupying new habitats in the vicinity of man, like unmown ditches alongside highways. These ditches correspond ecologically to the naturally inundated meadows in the lower reaches of rivers — they are flooded by water that runs off the road surface (including melting snow) and are supplied with nutrients washed down from adjacent fields. In such places foxtail forms border communities instead of flat spreading tracts.

The inflorescence is cylindrical and resembles a bushy tail, hence the scientific name derived from the Greek words *álópeks,* meaning fox, and *cýrá,* meaning tail.

Meadow Foxtail has a continuous distribution in all of western and central Europe, farther east in central Asia and by the large Siberian rivers.

It is a plural-yearly, stoloniferous grass with relatively short, conspicuously grooved leaves. The ligules are 2—5 mm long, collar-like and membranous.

The inflorescence is composed of a great many spikelets that are laterally compressed and with hairy glumes (1). Unlike the similar inflorescence of Cat's-tail, however, all the spikelets can be easily pulled off the stem (2).

Meadow Foxtail flowers early, from May to June.

Globe Flower
Trollius europaeus L.

Ranunculaceae

If it were not for the sporadic occurrence of *Trollius laxus* Salisb. on the North American continent, members of this genus could be described as Old World plants, not only geographically (prevalent in Eurasia) but also in time. The large, so-called globe-flower meadows found in practically every region in Europe from lowland to mountain districts are now almost a thing of the past, for the Globe Flower does not tolerate the chemicals of modern agriculture or thoughtless drainage. Its continuous distribution began in central Europe and its range extended to Great Britain but in western Europe it occurred only isolatedly. The latest taxonomic studies, however, led to the division of the original species into two: *Trollius europaeus* L.s.str. and *T. altissimus* Crantz. The first is widespread in Scandinavia, Finland, the European USSR, northern Germany (both the German Democratic Republic and German Federal Republic), Great Britain, Ireland, France, Spain, Switzerland and probably also north-western Italy. The second has a smaller range and is also much more variable; it grows in the German Federal Republic and German Democratic Republic, Poland, Czechoslovakia, Austria and in the Swiss Alps.

Globe Flowers are often grown in the garden. Besides the European, Asian and American species they are most often grown as cultivated hybrids, some of which are almost 80 cm high; double, and usually with orange or golden-yellow flowers.

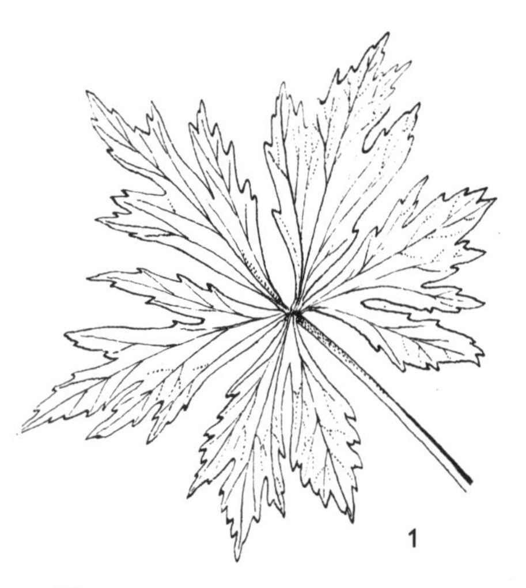

Trollius europaeus is a robust perennial herb with erect stem and alternate, deeply palmately-divided leaves (1). The flowers are hermaphroditic and regular; the unusual thing about them is that the five or more sepals are coloured yellow and serve as a conspicuous visual bait for insects. The corolla lobes, enclosed by the coloured calyx lobes, are small and narrow and have nectaries on the inner side. Striking is the difference between the broad calyx lobes and narrow corolla lobes in the Asian species *T. chinensis* Bunge.

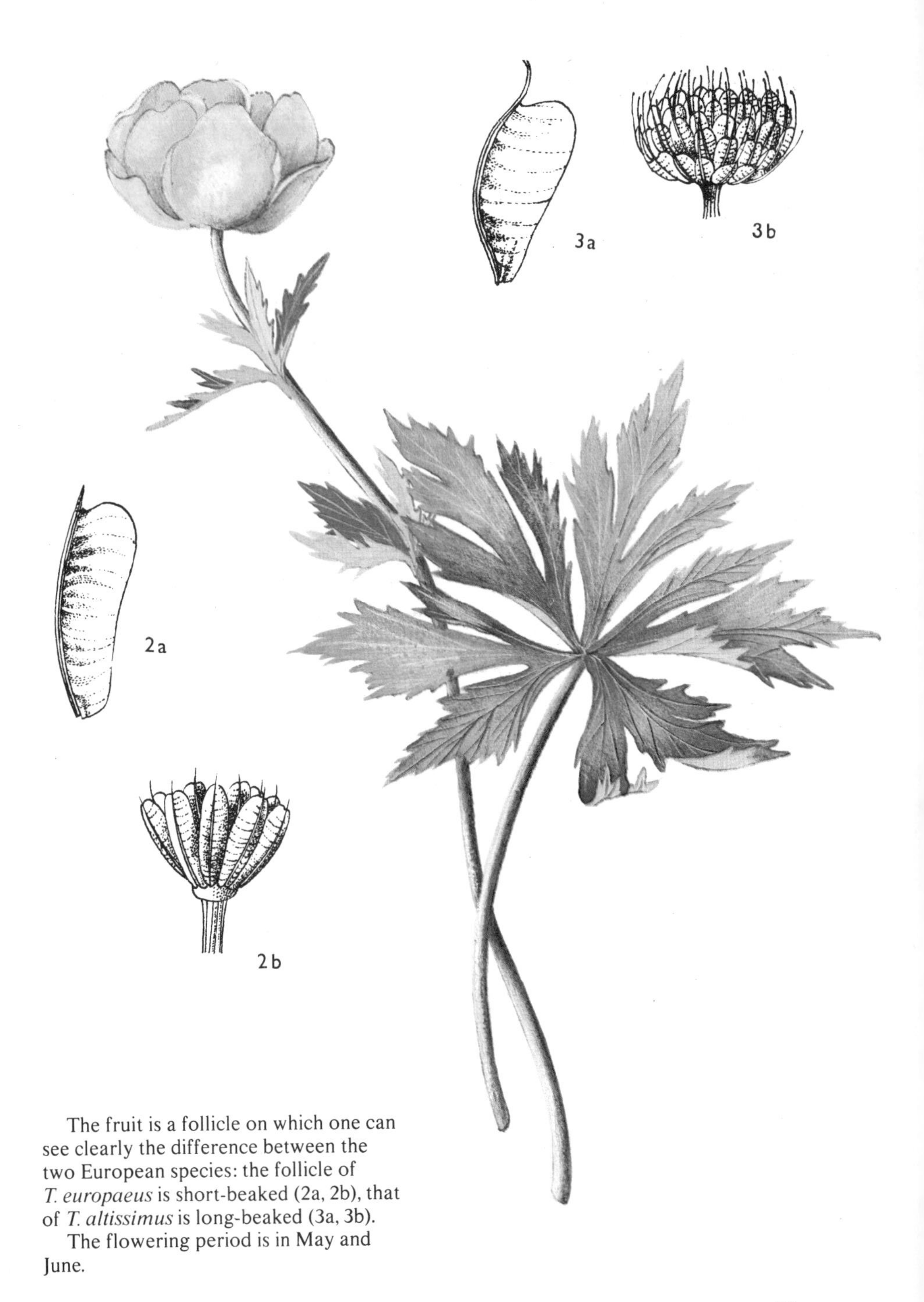

The fruit is a follicle on which one can see clearly the difference between the two European species: the follicle of *T. europaeus* is short-beaked (2a, 2b), that of *T. altissimus* is long-beaked (3a, 3b).

The flowering period is in May and June.

Ragged Robin
Lychnis flos-cuculi L.

Caryophyllaceae

Ragged Robin is one of the prettiest of European meadow plants with its bizarre flowers and glowing colours. Meadows which most of the year appear to be monotonous, flat, green tracts without vertical or structural dominants are transformed, seemingly overnight, in early summer. They are like vain ladies donning gowns of different colours every day — being at their loveliest, perhaps, in the rosy haze of the delicately-fringed blossoms of Ragged Robin.

From the scientific viewpoint Ragged Robin is a south-Siberian meadow geo-element found in western Siberia and practically all of Europe (even in Iceland). It has also been introduced to North America. It grows not only in damp and moss-moor meadows but also in scrub and waterside thickets as well as in fields as a weed. It grows and bears flowers equally well at both lowland and high mountain elevations (e.g. in Carinthia at approximately 2,000 m above sea level).

The other members of the genus *Lychnis* are also conservative inhabitants of the Old World. Indigenous to the warm slopes of the southern Alps, for example, is the one whose flowers were dedicated to Jupiter — *L. flos-jovis* (L.) Desr.; commonly grown in gardens is the fiery-red *L. chalcedonica* L. of eastern Europe and Siberia.

Ragged Robin is a robust, glabrous perennial herb with underground rhizome and 30- to 90-cm-high stems, which are sometimes sticky beneath the nodes as in the genus *Viscaria.* It is sparsely covered with opposite leaves, which are longish-ligulate and pointed at the tip. The tissues contain saponins.

The plant's most striking feature are the flowers. The petals (1) are rose-red (very occasionally white) and cleft almost half-way to the base into four narrow lobes. At the bend of the bifid claw is a corona composed of linear-bristly segments. The fruit is a globose-ovoid capsule.

The flowering period is from late May to early August.

Alsike Clover Fabaceae
Trifolium hybridum L.

Plants of the genus *Trifolium* are found on all continents, the greatest number in the Near East and Europe. Many species are cultivated plants and the damp-loving Alsike Clover is one of them. Some botanists consider it to be a separate, natural species, whereas others consider it a hybrid, a cross between Red Clover and White Clover. It has been grown a long time in Europe, where it is often called 'Swedish Clover'. If we accept the theory that it is a natural species then its area of distribution is located in central and eastern Europe; in Sweden and western Europe its spread in the vicinity of man dates from the 18th century.

From the economic viewpoint the yield of dry fodder from four mowings is approximately 200 quintals per hectare per year. In addition this species, like other meadow clovers, is an important food for bees, providing as much as 100 kg of honey per hectare.

Trifolium hybridum is a sparsely-branched perennial herb with long-stalked glabrous leaves. The flower heads are globose and composed of numerous flowers. These are usually dingy white, later pinkish and finally brownish when they are spent (1a). They have the characteristic structure of all flowers of the pea family (1b). The calyx of the usually flesh-red flowers of Red Clover (3) is conspicuously hairy on the teeth.

T.incarnatum L. (2), the clover more widely grown in southern and western Europe, differs from other clovers in the structure and size of the flower heads as well as in the colouring of the flowers.

The flowering period is from May onward, sometimes until September.

1 a
1 b
2

Meadowsweet

Rosaceae

Filipendula ulmaria (L.) Maxim.

Meadowsweet is a prominent damp-loving herb widespread in damp meadows, by the waterside and in thickets alongside brooks. According to these habitats (and also according to the composition of the vegetation found there) communities with *Filipendula ulmaria* are usually divided into two separate units: meadowsweet meadows, with Meadowsweet and Marsh Cranesbill prevailing, and meadowsweet border communities of slightly different floristic composition.

The top parts and flowers of Meadowsweet, characteristically smelling of methyl salicylate (oil of wintergreen), are used in the treatment of fever and also have a sweat-promoting and diuretic effect. Described as a 'plant salicylate', it is used in the treatment of flu and diseases of the urinary bladder. Spreading waterside masses of Meadowsweet produce large quantities of pollen and are often visited by bees.

Meadowsweet is distributed from western and north-western Europe (in Scandinavia as far as Nordkapp) to Siberia and Mongolia. The more thermophilous and xerophilous Dropwort (*F. vulgaris* Moench.) is distributed from western Siberia through the middle of the European continent to the British Isles.

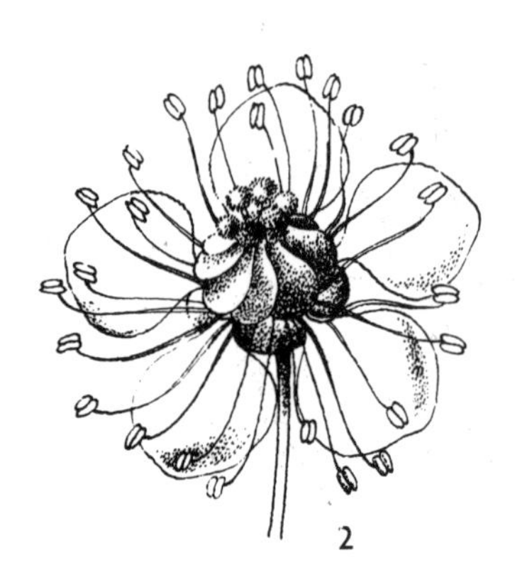

Meadowsweet (1) is a robust perennial herb approximately 1 — 1.5 m high with creeping rhizome and sparsely-branched stem. Characteristic of this plant are the compound leaves, which are odd-pinnate and composed of larger leaflets and conspicuously smaller interjected leaflets. The white flowers are usually five-petalled (2) and bloom in June, July and sometimes August. The fruit is a follicle.

The stems of Dropwort (3) are erect and practically leafless at the top. The compound leaves, odd-pinnate with interjected leaflets, are composed of a great many deeply incised leaflets. The white flowers generally have six petals.

1
3

Great Burnet
Sanguisorba officinalis L.

Rosaceae

Great Burnet is one of the less conspicuous meadow flowers for it is not brightly coloured. The flower heads are a dull crimson and one notices them only when they bloom in greater masses. Great Burnet was a common component of old, artificial, two- and three-crop meadows with a floristic composition similar to that of natural meadows. With the introduction of artificial fertilizers and the further chemization of agriculture, however, farmers noticed that Great Burnet was disappearing from meadows, slowly at first but then with increasing rapidity.

A like fate also befell those faithful meadow inhabitants — green grasshoppers and field grasshoppers — which were among the first to abandon such chemically treated tracts. In some places radical drainage likewise contributed to the disappearance of Great Burnet from damp meadows.

Members of the genus *Sanguisorba* are divided in their distribution on the continents of the northern hemisphere. The Great Burnet is distributed in western and central Europe to eastern Asia — as far as Alaska. In Scotland, Ireland and Scandinavia, however, it occurs only sporadically. The related Salad Burnet (*S. minor* Scop.) has a continuous distribution only in central and western Europe. North America, on the other hand, has several indigenous burnets — e.g. *S. canadensis* L. in the north-east and *S. sitchensis* C.A. Meay, *S. microcephala* Presl. and *S. occidentalis* Nutt. on the Pacific coast. Further burnets are to be found in south-east Asia.

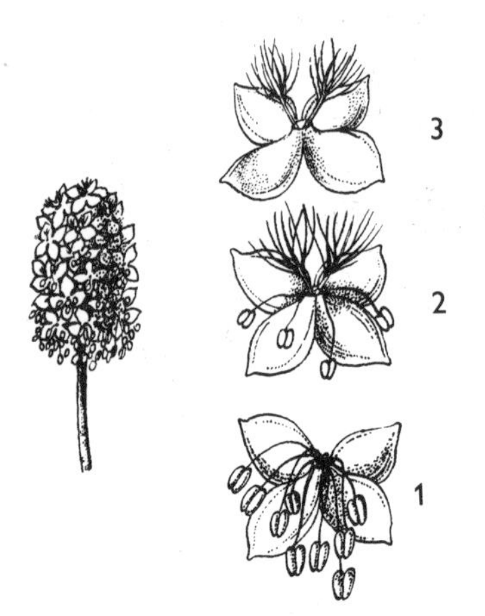

Sanguisorba officinalis, which translated from the Latin means 'medicinal cleanser of blood', is an ancient medicinal plant used with success since time immemorial to treat diarrhoea. It can be used also in baths and as a compress applied to suppurating wounds and skin rashes.

Great Burnet is a perennial herb with thick rhizome and erect, 30- to 90-cm-high stem branching at the top. The leaves, in a basal rosette, are long-stalked and odd-pinnate. The dark red flowers, clustered in heads, are 4-merous. They are without petals, the function of visual insect bait being served by the coloured calyx. All the flowers are

hermaphroditic and open from June to September.

The flower heads of Salad Burnet are more globose, green tinged with red, and composed of three types of flowers: male at the base (1), hermaphroditic in the middle (2), and female at the top (3).

Bistort

Polygonum bistorta L.

Polygonaceae

The pink spikes of Bistort are a common sight in damp meadows, on stream banks and around springs in lowland districts as well as in high-stemmed and alpine grasslands from mountain to sub-alpine elevations. In the Alps it is most widespread at elevations between 800 and 1,900 m above sea level but may be found even higher up. It is distributed throughout practically all Europe, particularly in mountain districts (in Scandinavia it is a naturalized plant), and through northern Asia (Siberia) to Kamchatka; it also grows in the arctic regions of North America, particularly Alaska.

The stout rhizome and top parts yield a medicinal drug used to treat inflammation and diarrhoea. Modern medicine, of course, views with humour the medieval use of the drug as a remedy for snake bites. The belief that it possessed such anti-toxic properties was based on the resemblance of the S-shaped root to the body of a snake. Hence, also, the name 'Snakeroot' by which it is known in some countries.

In Europe's extreme northern regions, as well as in practically all of north-east Asia and arctic America, grows the related Alpine Bistort (*P. viviparum* L.), which is sometimes classed together with *P. bistorta* in the separate genus *Bistorta.* A peculiarity of this northern mountain species is that it multiplies by means of bulbils that are sometimes formed instead of flowers.

Polygonum bistorta is a perennial herb with stout, twisted rhizome. This is reflected even in the plant's scientific name derived from the Latin words *bis* and *torquere,* meaning twice twisted. In competition with other high-stemmed herbs the stem of this polygonum reaches a height of 1 m. The basal, longish-ovate, pointed leaves have a conspicuously winged stalk; the upper stem leaves are faintly heart-shaped at the base.

The flowers are arranged in a striking terminal spike-like inflorescence up to 7 cm long. They are pink, occasionally whitish, with five perianth segments and eight stamens (1).

The root contains up to 20 per cent tannins, free gallic acid, catechin, starch, mucilage and other substances used in the preparation of tinctures for mouthwashes.

Sneezewort
Achillea ptarmica L.

Compositae

When the term *Achillea* is used it is usually envisaged as the Common Yarrow or Milfoil (*Achillea millefolium*) with leaves finely divided into some thousand leaflets (2), as its Latin name indicates. Few would be willing to believe at first glance that Sneezewort belongs to the same genus *(Achillea);* perhaps only the faint smell of the leaves when rubbed between the fingers calls to mind its kinship.

Species with narrow, undivided leaves, however, are not so rare amongst the members of the genus *Achillea!* Examples in Europe are *A. salicifolia* Bess. and *A. ptarmica* L. The first is found in waterside thickets, the second and more common species grows scattered in damp meadows and by water, in ditches and in reed beds. It is spread through the temperate regions of Europe, to the Arctic Circle, and in Asia (Siberia) — from lowland to median submontane and mountain elevations — in the Carpathians in moors and at elevations of approximately 1,000 m. However, in the Alps it is almost completely absent!

Sneezewort is one of the few plants of the genus *Achillea* that are grown in the garden; it is a traditional plant of country gardens and is grown chiefly for cutting (the flowers are long-lived).

Already in Pliny's day it was said that various species of this genus were used by Achilles' warriors to treat their wounds and ailments and that, supposedly, is how the genus came by its name — *Achillea.* Though this may or may not be true one thing is certain — they are very old medicinal plants.

3

Sneezewort is a perennial herb with a rhizome that becomes woody and an erect, densely-leaved stem, usually 30 — 150 cm high, branching in loose panicle-like fashion at the top. The stem leaves (1) are linear-lanceolate with serrate margin whereas those of other *Achillea* species are conspicuously divided (2). The flower heads, which are not large, are composed of 8, 10 or even as many as 15 white ray florets and creamy-yellow tubular disc florets (3). The outer ray florets are usually female, the disc florets hermaphroditic. The fruit is an achene (cypsela) without a pappus.

The flowering period is from July to September.

1
2

Marsh Thistle

Compositae

Cirsium palustre (L.) Scop.

The composite family appears to be a young, recent group on the evolutionary scale. The first fossil records were found in geological strata dating from the second half of the Tertiary period. Among them were the ancestors of present-day cirsiums.

The genus *Cirsium* is relatively large and includes some 250 species and a great many hybrids, for even such ecologically distant species as the xerophilous Dwarf Thistle (*C. acaule* Scop.) and the damp-loving Marsh Thistle interbreed. In such cases, however, the offspring is usually not very vigorous.

Marsh Thistle grows together with other high-stemmed herbs in damp meadows, valleys with brooks and streams, and round springs at the base of slopes — in other words in places with a high water table. Found at higher elevations in central Europe are communities of Marsh Thistle and Bistort that correspond to north-western Europe's more widespread communities of Smooth Brome (*Bromus racemosus* L.) and Marsh Ragwort (*Senecio aquaticus* Hill.).

Widespread in nitrogen-rich valley meadows and the herbaceous undergrowth of alder and aspen groves, often in the company of the same sort of herbs as Marsh Thistle, is the yellow-flowered Cabbage Thistle (*C. oleraceum* (L.) Scop.). These two, the Marsh and Cabbage Thistle, likewise often interbreed; their offspring, named *C.* x *hybridum* Koch, is quite common in the wild.

Marsh Thistle (1) is a 30- to 200-cm-high, spiny, biennial herb. The whole stem is covered with wing-like spines and in the bottom half with spreading hairs or gauzy-wool. The stem leaves are decurrent (the spiny wings on the stem appear to be a continuation of the leaves). The small heads of lilac-pink flowers are clustered on short branches at the tip of the stem.

The perennial Cabbage Thistle (2) is markedly different not only in the colour of the flowers, which are yellow, but also in having leaves that are soft, not prickly, fresh pale green and little divided or deeply lobed. Marsh Thistle flowers from July to October, Cabbage Thistle from June to October. They both provide food for bees and guarantee a relatively good yield of honey in mid- and late summer.

The fruit is a small achene (cypsela) with feathery pappus (3).

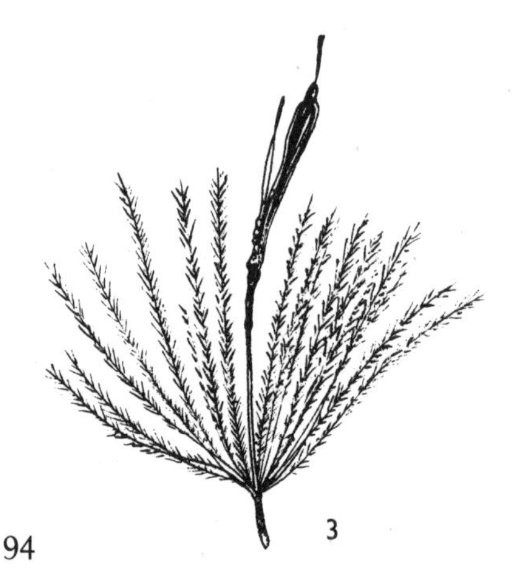

1
2

Marsh Marigold
Caltha palustris L.

Ranunculaceae

The Marsh Marigold — that golden-yellow herald of spring — has a widespread circumpolar distribution in the northern hemisphere, being absent only in the arctic regions of Canada and Greenland. It is actually a complex group of strains and varieties, a fact first noticed by Carl Linné himself. He observed that the various European populations had different flowering periods: March in Holland, from April to May in Sweden and June in Lapland. This most certainly has to do with the external ecological conditions of the respective regions but it has also led to various taxonomic and genetic conclusions. There is also a relatively marked variability in the external morphological characters, e.g. in the number of perianth segments, and so there may be some justification for the division of the species into smaller taxonomic units. Europe's distinguishable strains (subspecies), for example, were classed in separate systematic units according to the shape of the fruit.

The native type strain (*C. palustris* subsp. *palustris*) is most common in damp meadows with stagnant water or the shoreline zone round ponds, where it forms striking characteristically-textured masses.

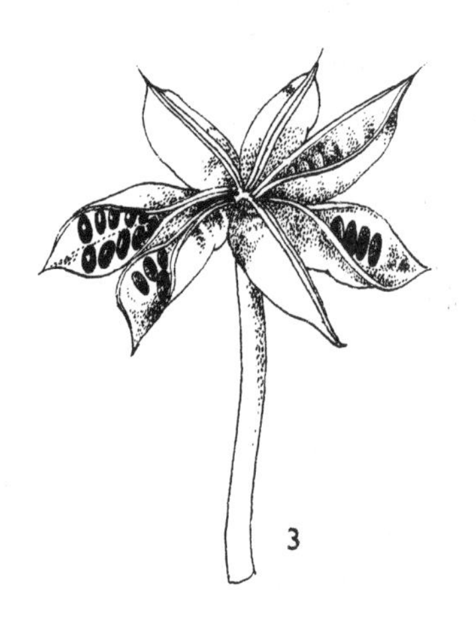

Marsh Marigold is a perennial herb with a fleshy, branched stem. In the type strain the stem is erect, in other strains it is prostrate and roots at the nodes. The basal leaves are orbicular with a heart-shaped base and long stalk (1), the upper leaves are sessile and nearly kidney-shaped (2). Those that appear later (in summer) differ from the spring leaves by being more coarsely and sharply toothed on the margin.

The flowers usually have five perianth segments and a great many stamens. The fruit is a follicle that remains on the receptacle a long time (3).

Because they contain a certain amount of alkaloids Marsh Marigolds are considered poisonous.

The flowering period is from March to June.

2
1

Marsh Valerian
Valeriana dioica L.

Valerianaceae

Marsh Valerian is the valerian family's representative in marsh and sedge meadows. The members of this family are distributed throughout the temperate zone of the northern hemisphere, chiefly in the Old World, and in South America.

Marsh Valerian grows scattered not only in marsh meadows but also in the grasses and reed beds bordering ponds, in damp ditches and in woodland wetlands — from lowland to mountain elevations (in Bavaria up to 1,000 m and in the Tyrol even at 1,700 m above sea level). Its continuous range of distribution extends from southern Sweden, Denmark, Great Britain and north-eastern Spain far into the USSR.

Valerians (*V. officinalis* L. and *V. dioica* L.) are important medicinal plants. They are grown in a great many forms to this day, particularly in Germany. The roots (rhizomes) contain numerous substances, chiefly essential oils, which have a sedative effect on neurogenic irritability and beneficial action on heart disorders of nervous origin and are used also as a general sedative in insomnia caused by nervous exhaustion, increased heart rate and agitation. The drug is usually prescribed as a tincture *(Tinctura valerianae)*. Valerians are plants with an aromatic odour — caused, among other things, by isovaleric acid — which attracts cats.

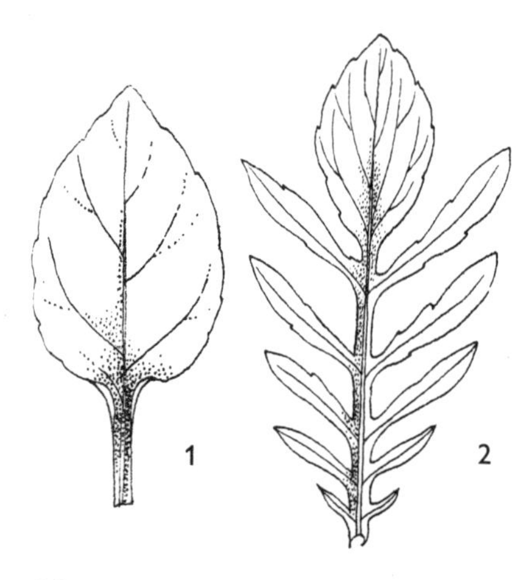

Marsh Valerian is a glabrous perennial herb, barely 10—30 cm high, with short, creeping rhizome and erect, sparsely-leaved stems. The basal leaves are usually undivided (1), the remaining stem leaves are sessile and pinnate with larger, ovate terminal leaflets (2). The flowers are borne in three-branched terminal dichasiums. Male plants have larger, usually pink flowers with three

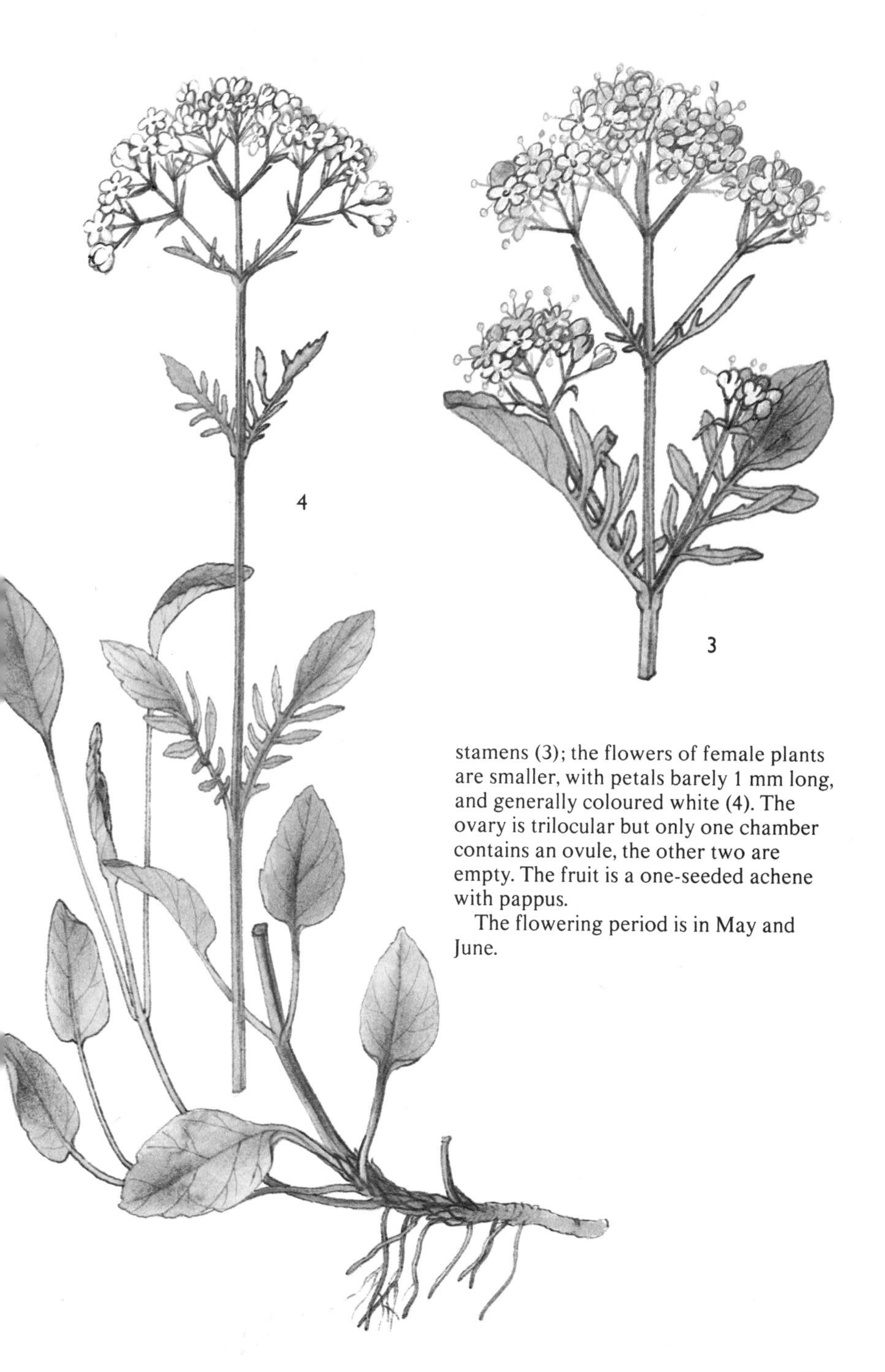

stamens (3); the flowers of female plants are smaller, with petals barely 1 mm long, and generally coloured white (4). The ovary is trilocular but only one chamber contains an ovule, the other two are empty. The fruit is a one-seeded achene with pappus.

The flowering period is in May and June.

Marsh Cinquefoil
Potentilla palustris L.

Rosaceae

If plants of bogs and marsh meadows were to compete as to see who was most elegant it would be likely that Marsh Cinquefoil would walk off with the prize. However, amidst the masses of grasses and other plants of marsh meadows and in the undergrowth of waterside willow groves it readily escapes notice for its soberly coloured flowers are superbly camouflaged.

Inasmuch as the ecological conditions of freshwater moors in the northern temperate zone are practically the same in Europe, Asia as well as North America it is not surprising that the Marsh Cinquefoil is a circumpolar species found in all the aforesaid continents. It occurs from lowland to relatively high mountain elevations — in the Alps at altitudes of approximately 2,000 m above sea level. It is most commonly found in depressions in marsh meadows, near smaller bodies of water, together with Marsh Trefoil, Cotton-grass and sedges.

Marsh Cinquefoil differs slightly from others of the genus *Potentilla* in the structure of the flowers and many botanists therefore class it in a separate genus — *Comarum.*

The flowers of Potentillas have an epicalyx and a slightly domed receptacle with greater number of pistils arranged in a spiral. The 5-merous flower of Marsh Cinquefoil (1) — the terminal flower is often 7-merous — has an atypical structure consisting of an epicalyx with narrow, lanceolate segments coloured green, a calyx with broadly ovate pointed segments coloured green outside but dark

purple to brown within, small petals, only 3—8 mm long, coloured red or purple and a great many stamens (about 20) likewise coloured dark purple.

The flowering period is in June to July.

The fruits are achenes arranged on the raised receptacle in a spiral. The entire aggregate fruit (head of achenes (2)) resembles the dry globose berry of the Mediterranean shrub *Arbutus unedo* L.

Purple Loosestrife
Lythrum salicaria L.

The Lythraceae are a remarkable family of plants. Their fossil records found on the territory of present-day England date from the early Tertiary period. They include some 25 genera and 450 species found in damp places or in water from tropical to temperate regions. The species *Lawsonia inermis* of Iran, cultivated in tropical Asia and north Africa, yields a product known to women the world over — namely the reddish-brown dye henna. It is extracted from the leaves and has been used to tint hair auburn since the days of the early Egyptians.

Purple Loosestrife, found also by the waterside and in wet marsh meadows, grows equally well in the Hebrides and Scandinavia, on the Kola Peninsula, round Lake Baikal in Siberia, the mouth of the Yangtse River in China, in Tibet, round the River Jordan, in Algeria, Canada, Peru and even south-eastern Australia.

It became part of the history of modern natural science thanks to Charles Darwin's work 'The Different Forms of Flowers on Plants of the Same Species' (published in 1877) and became a model used to explain the condition known as heterostyly in the plant realm and to substantiate its purpose as a mechanism preventing self-pollination and encouraging cross-pollination.

In folk medicine the leaves of Purple Loosestrife were used as an agent arresting the growth or multiplication of bacteria — country folk applied the freshly-plucked leaves to cuts.

Purple Loosestrife is a perennial herb 30—150 cm high with a cephaloid rhizome and erect robust stem that is sharply four-angled and branches from the base. The leaves are narrowly-lanceolate with entire margin. The flowers are usually arranged in a spike-like inflorescence at the tip of the main stem or side branch. The scientific name of the genus is derived from the Greek expression *lythron*, meaning soiled by blood.

Masses of Loosestrife include plants with three types of flowers: the first has a long style, extending far beyond the six short and six longer stamens with yellow anthers (1); the second has a very short style, six longer stamens with greenish-blue anthers and six shorter stamens with yellow anthers (2); the third has styles of median length, six longer stamens with blue anthers and six shorter ones with yellow anthers (3). Stigmas are fertilized only with pollen from stamens of the same length as the styles. The flowering period is from July to September.

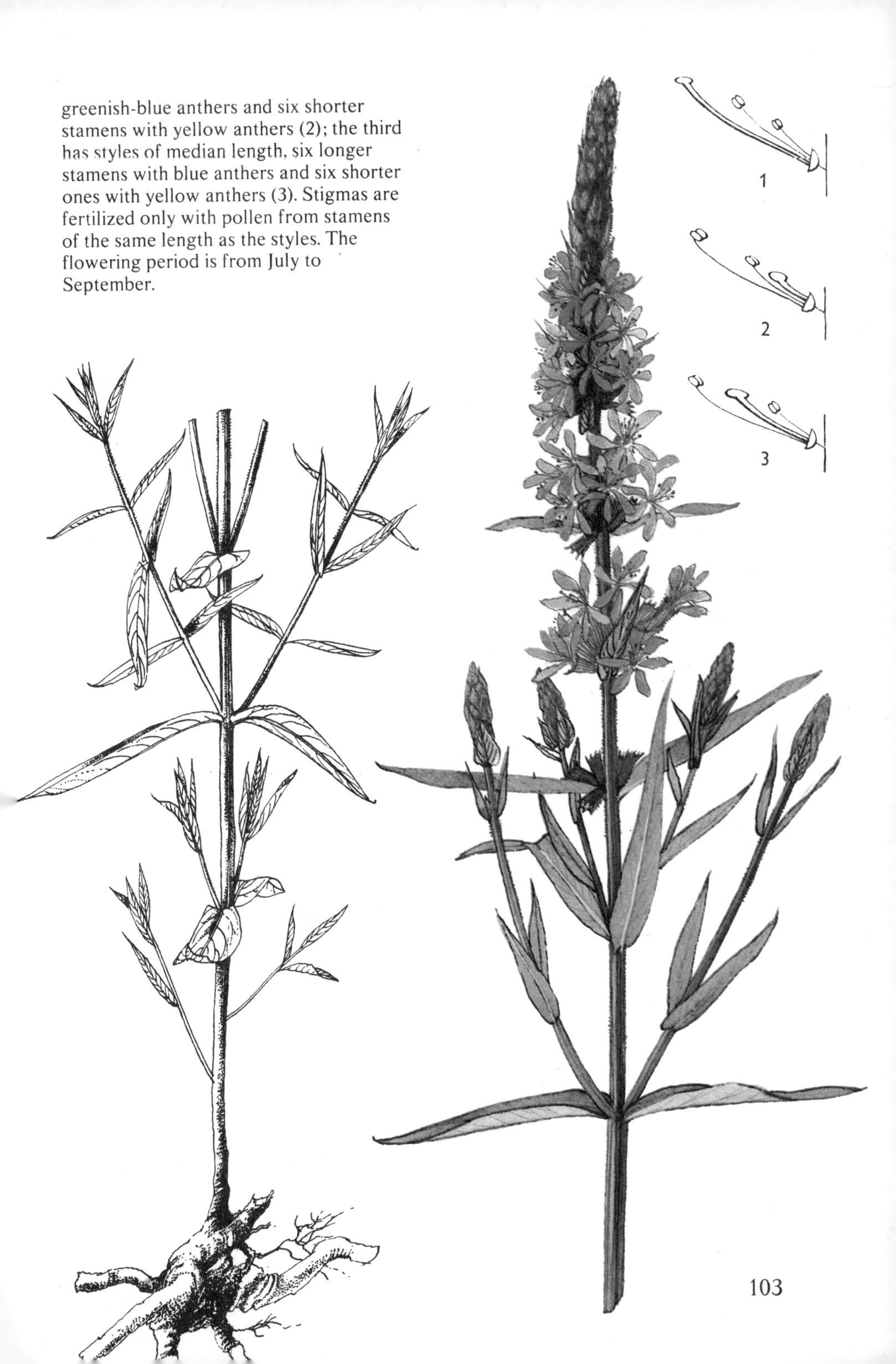

Common Sundew
Drosera rotundifolia L.

Droseraceae

The sundew, a 'carnivorous' plant, is a sort of tropical token — there is no scarcity of large 'carnivorous' plants in the tropics — which grows in the temperate regions of the northern hemisphere and in soils that are very deficient in nitrogen and minerals. They have survived in such uncongenial places as heather moors, moss moors, damp meadows, ditches and watersides because they obtain the substances that are in short supply (nitrogen, phosphorus, potassium) from the bodies of insects dissolved by their secretions. The leaves of sundews are covered with sensitive hairs called tentacles which have drops of sticky liquid at the tip; the tentacles also contain vascular bundles providing for the transport of the nutrients obtained from the bodies of the insects. The initial phase in capturing an insect is a passive one — the insect alights on the leaf and is held fast by the sticky liquid. In the second phase the plant plays an active role — in response to a chemical stimulus or to mechanical irritation caused by the weight of the insect the leaf begins to grow asymmetrically and to close round the insect so that its body comes in contact with a greater number of tentacles.

D. rotundifolia and related species, e. g. Great Sundew (*D. anglica* Huds.) are so-called boreal-circumpolar species. The first is distributed from 30° latitude North to the Arctic Circle in Eurasia and North America. The main area of distribution of the second is in north-western and northern Europe, Siberia, Sakhalin, North America and Canada; it is found in high mountains such as the Pyrenees and Alps and also in more southerly regions. From as far back as medieval days sundews were used as an effective remedy for sclerosis; this, however, led to the total extinction of these plants in numerous localities.

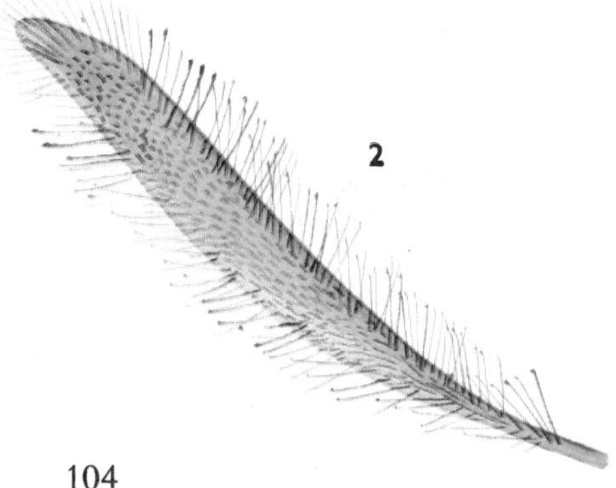

Common Sundew is one of 15 species of *Drosera* found mostly in Australia, New Zealand, south Africa and Brazil. The scientific name of the genus is derived from the Greek word *drósos*, meaning dew.

Sundews are perennial plants with a basal rosette of leaves, orbicular in the case of Common Sundew (1), linear-oblong in Great Sundew (2). The green leaves are literally covered with numerous gland-tipped hairs (tentacles).

The flower stem is terminated by a cyme composed of tiny white flowers that blossom in July and August but remain open only a few hours. The fruit is a dehiscent capsule. The seeds are minute and light and thus readily distributed by both wind and water, sometimes being transported great distances and to otherwise isolated places.

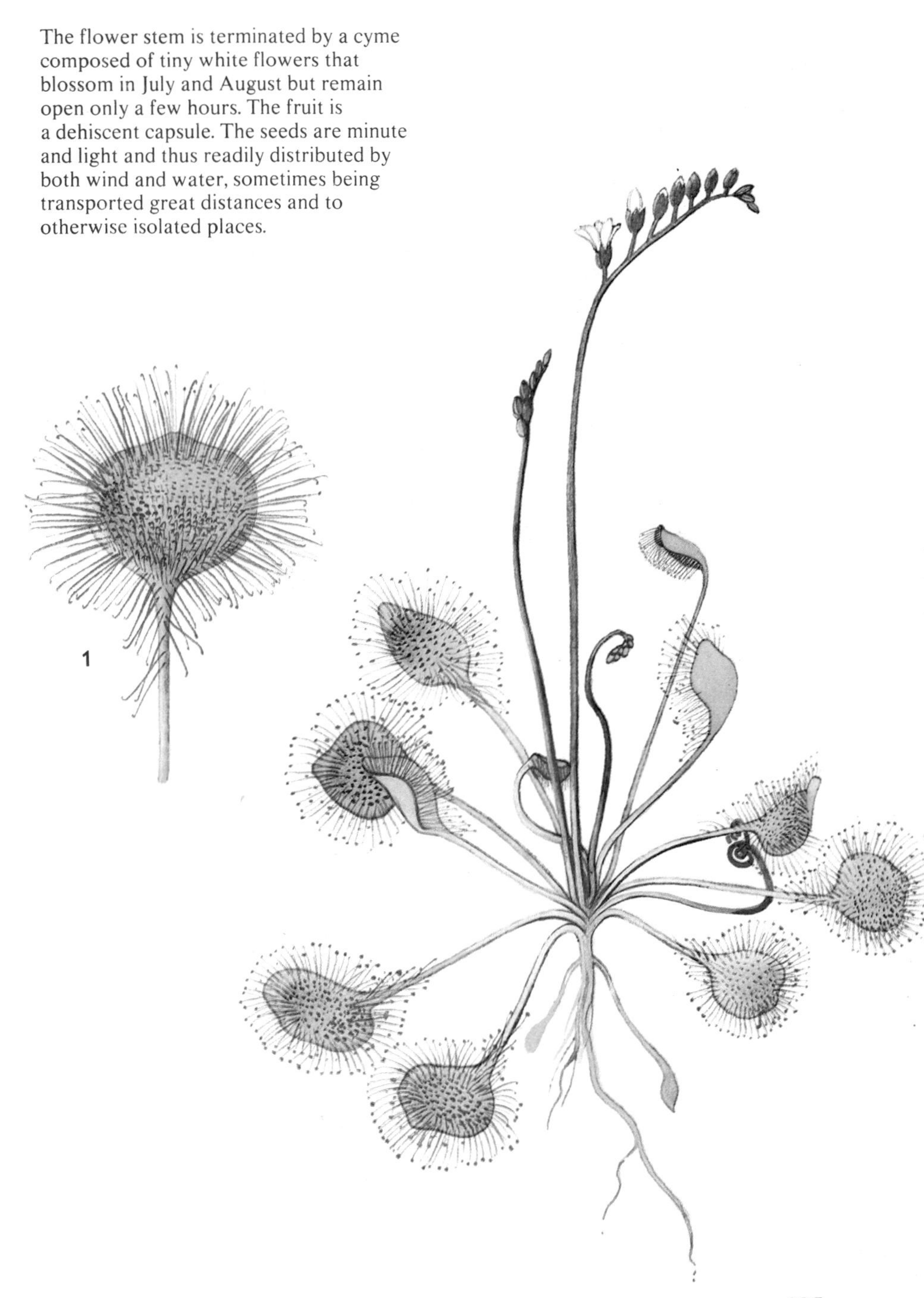

Grass of Parnassus Parnassiaceae
Parnassia palustris L.

Grass of Parnassus is a plant that is unusual and unique in many ways. Its very classification poses a problem. Many botanists even considered it to be closely related to sundews. Usually, however, members of the Parnassus grass family (Parnassiaceae) are classed as a separate family in the order Saxifragales, a family that is monotypic and consists only of a single genus — *Parnassia* — embracing some 45 amphiboreal species distributed in the north temperate zone.

Grass of Parnassus has a continuous range that includes all of Europe, excepting certain Mediterranean regions, and extends through northern Asia (Siberia) to the Far East and Japan; in North America it occurs chiefly in Canada. It avoids the arctic regions and Greenland but is found in Iceland.

It grows in damp and marsh meadows and moors in lowland districts; in high mountains it grows on limestone rocks often at subalpine and alpine elevations (in the Swiss Alps up to 2,700 m above sea level). Despite this great ecological amplitude, particulary its tolerance as regards soil acidity, Grass of Parnassus is a relatively stable species exhibiting little variability.

Formerly it was used in folk medicine to stimulate the appetite (the leaves are slightly bitter) and as a cardiac and eye medicine.

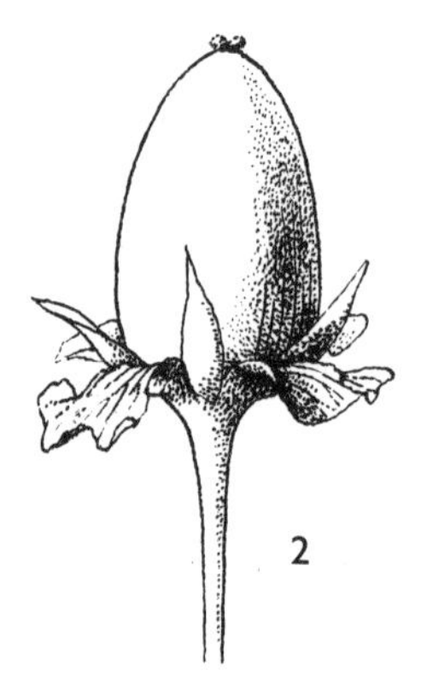

Parnassia palustris is a perennial herb with basal rosette of long-stalked leaves and erect stem, sometimes only 5 cm and at other times up to 40 cm high, that has only a single leaf and is terminated by a solitary flower. This leaf is stalkless and usually located at a point about two-thirds up the stem. Some plants have practically no basal leaves and a mass of such plants then consists only of single stems with clasping leaf.

The flowers, which remain open as long as eight days, are relatively large (even 3 cm in diameter) and 5-merous.

The petals are usually white and, very occasionally, tinted red. There are five stamens and five oblong staminodes (1) that serve as nectaries. The flowering period is from July to September (in mountains as much as a month earlier). The fruit is a capsule (2) that opens by four valves.

Marsh Gentian

Gentianaceae

Gentiana pneumonanthe L.

'Plant that alleviates lung diseases', such was the old saying pertaining to Marsh Gentian, which was reflected also in its scientific name derived from the Greek words *pneumón,* meaning lungs, and *anthós,* meaning flower. However, the large amount of bitter principles they contain earmarked many gentians for use in the preparation of bitter beverages and aperitifs by the food industry. And, inasmuch as the best drug is obtained from the underground parts (rhizomes), many herb healers of past centuries as well as the present have been responsible for the inclusion of gentians in practically every country's list of endangered and rigidly protected plant species (Red Data Book).

Marsh Gentian is slightly better off than the Yellow Gentian and Dotted Gentian because it was not collected so intensively, but still it now grows only in damp meadows, heather moors and moss moors as well as (thanks to its ecological adaptability) in rather dry semi-steppe meadows. It does not tolerate the application of fertilizer and intensive cultivation.

It is distributed throughout practically all Europe — northward almost to 60° latitude North — and western Asia. Aesthetically it is one of the loveliest of the high-stemmed gentians: a single plant may bear tens of flowers. Observations revealed that these respond in an unusual manner to irritation by insects (the corolla tube closes slightly).

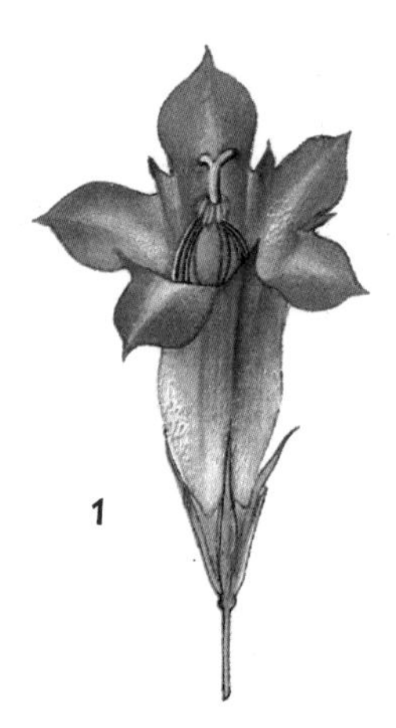

Gentians are royal plants. This word is perhaps applied to them because the genus was purportedly named after Gentis, King of Illyria — but the more likely reason is their appearance.

Marsh Gentian is a perennial herb 20—50 cm high with thick, sometimes polycephalous rhizome. The stems are erect, rigid, branched and bluntly angled. The leaves are sessile, opposite and linear-lanceolate in outline with slightly revolute margin. The flowers grow from the axils of the uppermost leaves (one to three in each axil). The funnel-shaped corolla (1) is much longer than the lobed calyx and coloured deep blue with five green-spotted lines on the outside.

The flowering period is from July to September (October) and that is why this Gentian goes by the name 'Autumn Bellflower' in old herbals.

The fruit is a capsule.

Broad-leaved Cotton-grass Cyperaceae
Eriophorum latifolium Hoppe

Only few plants can boast that their greatest ornament is not the newly-opened but the spent flowers. Cotton-grass is definitely one of those few; however, the tuft of white bristles (the 'ornament') has a different purpose. Cotton-grasses often grow in windy open spaces and it is the wind that carries off and disperses the white tufts together with the mature achenes.

Cotton-grasses have a large area of distribution: Common Cotton-grass (*E. angustifolium* Honck.) and Hare's-tail Cotton-grass (*E. vaginatum* L.) are circumpolar species found in the north temperate zone including the arctic regions; Broad-leaved Cotton-grass is a Euro-Siberian element found in the temperate regions of the Old World.

Cotton-grasses are readily identified from afar in their native habitats — springs, moss moors, marsh meadows and pond margins — where, faced with little competition, they can form spreading masses that are a breathtaking sight in summer. Mountain moors, from piedmont to sub-alpine elevations, are more often inhabited by Hare's-tail Cotton-grass which differs from the others by bearing only a single spike at the tip of the stem. In the central Alps it grows at elevations of approximately 2,600 m.

The distinctive characteristic of all cotton-grasses is reflected in the generic name *Eriophorum*, which is derived from the Greek words *érion*, meaning wool, and *férein*, meaning to carry.

Cotton-grasses are perennial herbs which may be tufted (e.g. *E. latifolium* (1)), stoloniferous, or loosely-tufted (*E. angustifolium*). The erect stems are terminated by a single (*E. vaginatum* (2)) or several, long-stalked, later drooping spikes (3) composed of a great many flowers. The perianth segments are modified into smooth bristles which become longer after flowering and form a white tuft (4).

Cotton-grasses flower relatively early (from March to June, depending on the locality) but the flowers are not striking (5). Only later, after the perianth bristles have lengthened, do they come into their own, brightening meadow and shoreline with their white cottony heads.

Large Bitter-cress
Cardamine amara L.

Cruciferae

If you like to go hiking here's a good tip for a meal made tasty by nature's bounty. In spring when fresh vegetables are scarce take just a slice of bread and butter or cheese along with you and head for the boggy margin of a stream or spring. There you will find the white-flowering Large Bitter-cress (*Cardamine amara*). Do not let yourself be put off by the word 'bitter' and pluck some leaves to put on your bread. You will be surprised by the pleasantly spicy, slightly burning taste. If you take some leaves home you can use them to flavour a lettuce salad or similar dish.

Some members of this genus (*Cardamine*) have an unusual method of vegetative (asexual) propagation. Small buds that may develop into new plants form on the underside of the leaves at the point where the leaflets join the rachis. If the parent plant is flattened to the ground (e.g. by flood water) and the leaves come into contact with the soil a single leaf may give rise to several new plants. This characteristic is observed most often in Lady's Smock (*C. pratensis* L.).

Large Bitter-cress is distributed throughout Europe, in Scandinavia to 64° 30′ latitude North, in the south to the Apennines and the northern Balkans. In some places it forms spreading carpets alongside brooks and streams, by springs and in alder groves — from lowland districts to mountain valleys at altitudes of more than 2,300 m.

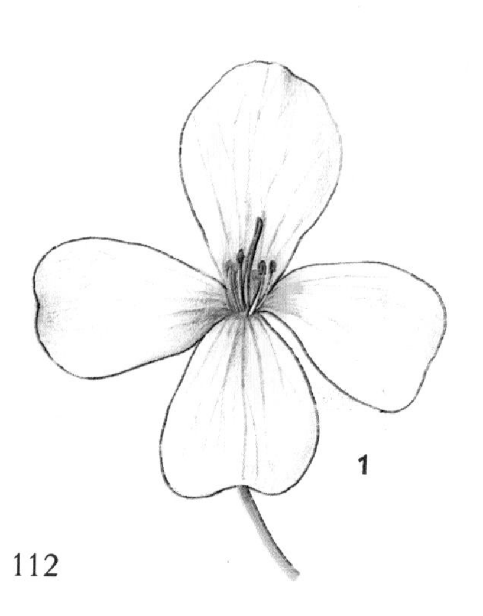

Large Bitter-cress is a perennial herb with creeping rhizome and rigid, pith-filled stem, 15—50 cm high. Prostrate stems readily root at the nodes in moist soil. The leaves are never arranged in a ground rosette. The stem leaves are odd-pinnate and broadly ovate; the terminal leaflet is conspicuously larger.

The flowers (1) are almost always white, only occasionally are they pale violet (with violet veins). The anthers, on the other hand, are a striking violet, a characteristic that makes it easy to distinguish this plant from the similar Common Watercress (*Nasturtium officinale*). The fruit is a straight siliqua with seeds arranged in a single row.

The flowering period is from April to June. When the seeds are mature (in full summer) the plant sometimes dies down, i.e. the top parts wilt and dry up.

Common Butterbur
Petasites hybridus (L.) G.M.&Sch.

Compositae

Butterburs are unique plants in many ways. Their systematic classification and their methods of propagation, for instance, have confronted scientists with many problems.

Come early spring the first butterburs begin cropping up on stream banks, chiefly in foothill and mountain districts. During the first warm days scaly stems terminated by racemes of scanty flower-heads rise from the firm base. On one plant the heads are either composed of hermaphroditic flowers in the centre and only a few pistillate flowers (or none) on the margin or else of only a few hermaphroditic flowers in the centre and a great many pistillate flowers on the margin. The first type are considered to be staminate (male) flowers because of their large production of pollen, the second pistillate (female) flowers. In nature, however, the occurrence of the two 'sexes' is not uniform. In Great Britain, for instance, 'female' plants occur more often only in middle England whereas 'male' plants are to be found everywhere. And yet practically no place is without butterburs, for they have the remarkable ability of spreading rapidly by means of long underground rhizomes. That is why you will rarely come across only a single plant in the wild; it is found usually in a large spreading mass. The flowering period is followed by the second stage in the life of the butterbur — the appearance of its large leaves. They are among the largest found in the temperate zone — one Common Butterbur, for example, had a leaf that measured 1.20 m by 1 m!

Butterburs have also been effective drug plants for centuries. The rhizome yields a drug used to treat diseases of the respiratory passages and to suppress the cough reflex.

Common Butterbur has flowering stems with red scales and short stigmas (1). The leaves are green both above and below, regularly toothed on the margin and heart-shaped at the base. It grows in lowland as well as mountain districts throughout Europe. The related White Butterbur (*P. albus* (L.) Gaertn.), found in foothills and mountains particularly in the sub-oceanic part of Europe, has flowering stems with scales coloured pale green and long-pointed stigmas (2). The leaves are white-felted beneath, deeply double-dentate, and with lobes practically touching at the base (3). In recent years western Europe has witnessed the rapid spread (by vegetative means) of *P. japonicus* F. Schmidt (4), brought there for garden cultivation from eastern Asia.

Butterburs flower in April or early May.

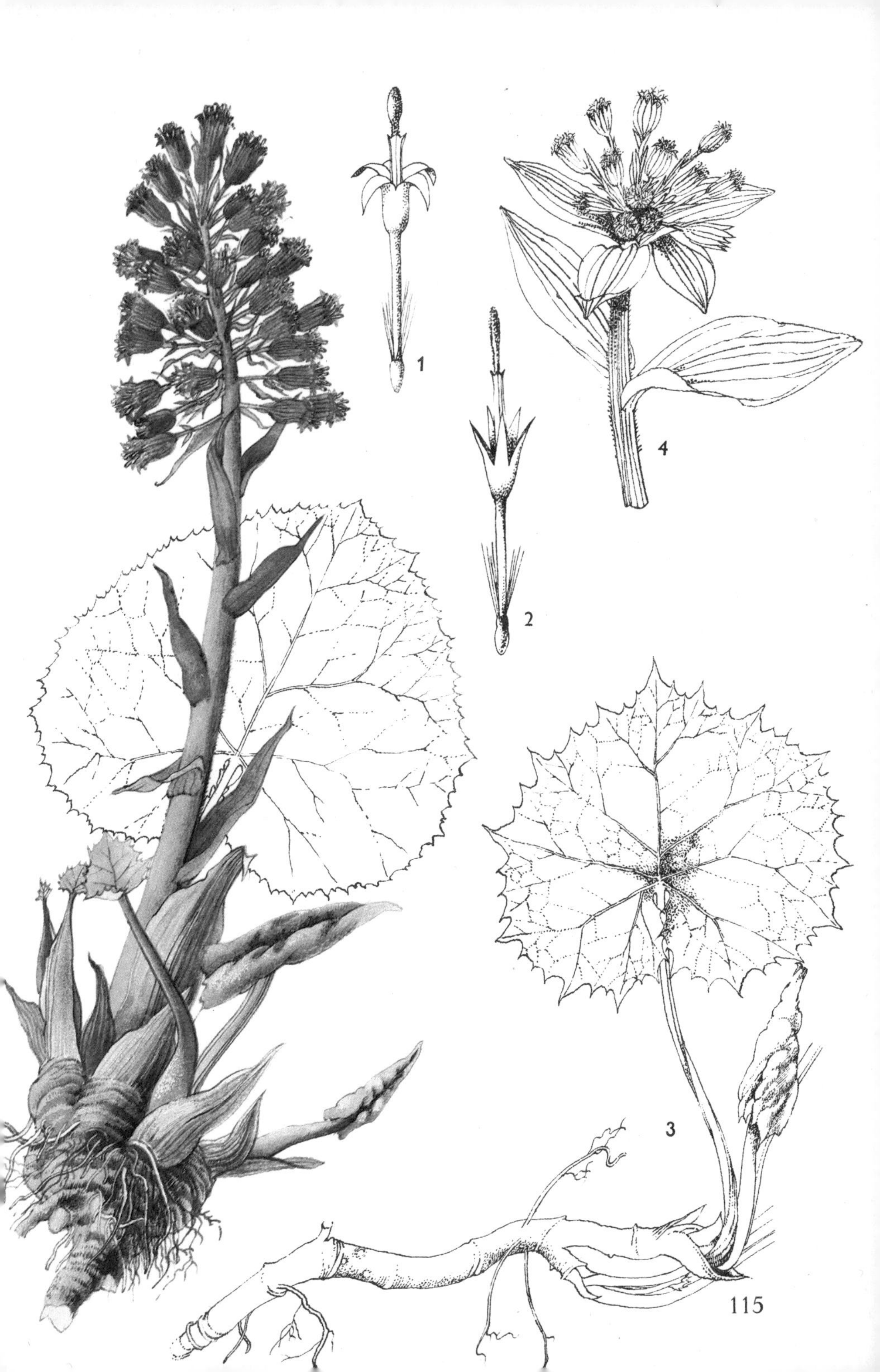
1
2
3
4

Bogbean or Marsh Trefoil
Menyanthes trifoliata L.

Menyanthaceae

Like the closely related gentians, members of the bogbean family have a bitter flavour. Because their leaves resemble those of trefoil they are often called 'Bitter Trefoil' in many languages, or else 'Water Trefoil' in reference to where they grow. All plants of the bogbean family grow in bogs and water in tropical as well as temperate regions.

Bogbean, a relatively rare herb nowadays, is distributed throughout practically all of Europe, including Iceland, the temperate regions of Asia to Japan and in the more northerly parts of North America. It grows scattered in bog and marsh meadows, often also in pools encroached upon by vegetation and at the edges of ponds in the sedge zone. It is a pioneer plant in the colonization (filling in) of shallow reservoirs.

Because of the bitter principles it contains Bogbean was used in folk medicine in place of Centaury where possible — chiefly in the treatment of gastric and digestive ailments. It was even added to beer to enhance its pleasant, bitter taste. The dried leaves are used by the pharmaceutical industry to this day for the effective substances stimulate the flow of gastric juices; they are also used in the preparation of bitter herb liqueurs. Because of its relative paucity in the wild Bogbean has been put on the list of endangered species and is being cultivated on bogbean plantations for pharmaceutical purposes.

Bogbean is a relatively small (15- to 30-cm-high) perennial herb with a creeping, jointed rhizome. The scientific name of the genus was first used by Theophrastus and apparently is associated with the striking structure of the flower (The Greek word *menýein* means to appear and the word *ánthos* means flower). The leaves growing from the rhizome have thickened sheaths and are composed of three leaflets at the end of a long stalk. They contain the glycosidic bitter principle loganin, tannins

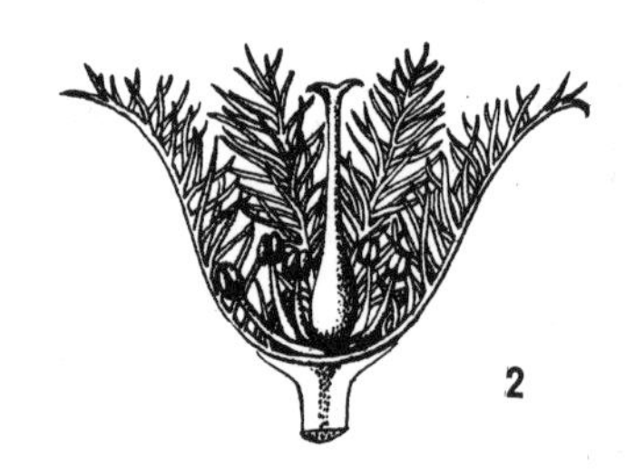

and other substances. The stems are usually leafless and terminated by a raceme of decorative, whitish flowers with corolla divided into fringed lobes. The flowers are usually heterostylic with styles either shorter (1) or longer (2) than the stamens. The fruit is a capsule.

The flowering period is in May and June.

Tufted Loosestrife
Naumburgia thyrsiflora (L.) Duby

Primulaceae

If recognized as a separate unit then *Naumburgia* is a genus with only a single species — Tufted Loosestrife. Small it may be, but its distribution is extensive. Tufted Loosestrife is a circumpolar species found in Europe, northern Asia and North America. It grows in reed beds, waterside thickets (chiefly willow thickets) and alder woods, generally on the borderline between continuous stands of vegetation and open water — in other words in places with full sun, though it also grows in the undergrowth of dense, shady willow and alder groves. As a result of its ecological adaptability Tufted Loosestrife occurs in three forms: terrestrial, littoral and submerged. It also stands up well to cold weather and its range extends practically to 70° latitude North.

Tufted Loosestrife is often included (according to Linné) in the genus *Lysimachia* as *Lysimachia thyrsiflora* L. However, it differs from the other members of that genus (see page 60) chiefly in the structure of the flowers, which are 6- to 7-merous (those of *Lysimachia* are 5-merous) and have narrowly linear petals; the flowers are arranged in dense stalked racemes that grow from the axils of the leaves.

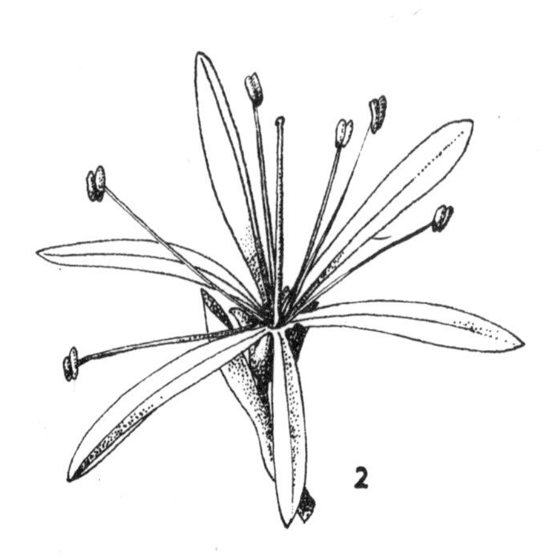

Naumburgia thyrsiflora is a perennial herb with creeping, jointed rhizome and underground shoots bearing reproductive buds whereby it spreads by vegetative means, often forming large continuous masses with a common root system. The erect stem is hollow, 30—60 cm high and scaly at the base (1). The leaves are sessile, opposite or in whorls, narrowly-lanceolate in outline, and slightly revolute on the margin. They are usually thickly dotted with red, glabrous on the upperside and woolly-pubescent beneath. The flowers have a five-lobed calyx and corolla usually 6- to 7-lobed (2) and flower from May to July. The fruit is a capsule that opens by five valves.

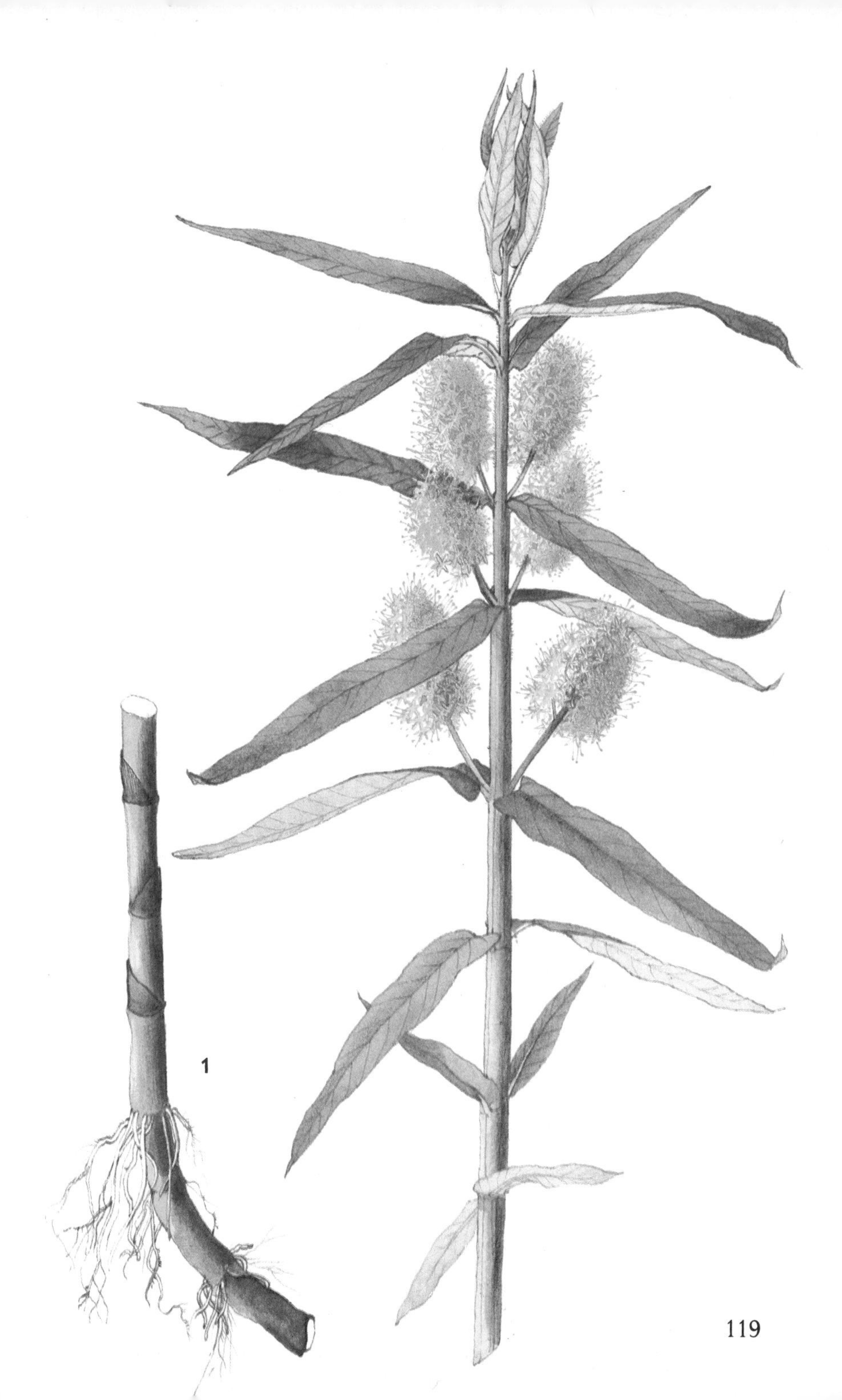
1

Wood Club-rush
Scirpus sylvaticus L.

Cyperaceae

This club-rush bears the Latin name *sylvaticus* meaning woodland — but in woods it generally grows in damp rides in hollows and at the edges of puddles and wheel tracks where it has been brought by man. In natural woodlands, however, it occurs only in damp alder stands. Otherwise it generally grows in so-called marsh-marigold meadows where it is often the prevailing species and forms whole separate communities on stream-banks, in ground depressions and round springs. In such places, as a rule, spreads of Wood Club-rush are accompanied in the main by sedges, buttercups and rushes and are relatively dependent on the stagnant water table. Where the underground water flows more rapidly beneath springs Wood Club-rush grows in communities poor in diversity of species together with Floating Sweet-grass (*Glyceria fluitans* (L.) R. Br.)

Spreads of Wood Club-rush are usually quite thick, covering 90 to 100 per cent of the ground. Their economic value is negligible; club-rush was formerly used to weave matting and the like (the same as reedmace, reeds and rushes) and as bedding for livestock.

It is a plant of hilly country and piedmont; in the Alps it was found growing at elevations of approximately 1,800 m. It has a continuous distribution that embraces practically all of Europe, excepting the Mediterranean and arctic regions, and is distributed intermittently also in Siberia.

Wood Club-rush is a 40- to 100-cm-high perennial herb with stoloniferous rhizome and erect, bluntly three-angled, hollow stems. The leaves are about 1 cm wide, keeled, and rough on the margin. The stems are terminated by a strikingly large inflorescence (anthella) — sometimes up to 30 cm long! — composed of multiflowered spikelets borne at the tips of the branches. The flowers in the spikelets are hermaphroditic with six characteristic perianth segments resembling bristles, three stamens, and a trifid stigma (1).

The flowering period is from May to July.

Bladder Sedge
Carex vesicaria L.

Cyperaceae

Selecting a suitable example of the genus *Carex*, one that would be representative of the group, is not easy, for this genus is one of the largest in the plant realm. It embraces some 2,000 species found practically throughout the whole world. The number of central European sedges alone is nearly a hundred! It is a genus of plants of extraordinarily vast ecological amplitude. Included among their number are sedges from dry, rocky, limestone slopes as well as large, tufted herbs that are characteristic features of wetlands in almost all parts of the world.

Sedges by the waterside and in damp meadows are not a welcome sight to fishermen and farmers. In shallow ponds tufts of sedge presage the entire filling-in of the pond area and their toughness and acidity make them unsuitable as fodder. Young spreads of sedge used to be grazed only by horses. Dense spreads of Alpine Sedge (*C. brizoides* L.), growing in damp woods and on stream banks, used to be cut and dried for upholstery padding.

The rhizomes of some stoloniferous sedges, e.g. *C. gracilis* Curt., exhibit interesting geotropical reactions: the rhizome always grows horizontally (at right angles to the earth's gravitational pull) no matter which way the plant is turned.

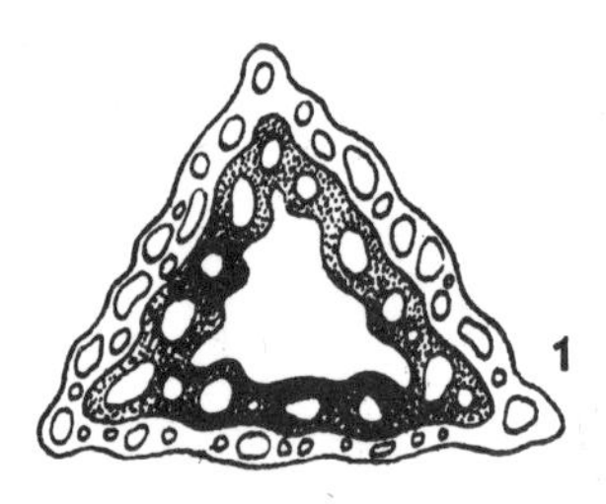

Sedges are perennial herbs, some tufted, others, such as *Carex vesicaria* and *C. gracilis*, with creeping (stoloniferous) rhizomes. Masses of such sedges look very much like grassy meadows; however, unlike grasses, many sedges have a conspicuously angled (usually three-angled) stem (1).

Both the sedges selected have an inflorescence composed of male spikelets (at the top) and female spikelets (at the bottom). Bladder Sedge (2) is one of the commonest shoreline sedges with striking, yellowish-green inflated follicles (3).

The greyish-green, stoloniferous *C. gracilis* Curt. (4) is also a common and widespread sedge found from lowland to high mountain elevations. The follicles (5) do not have a prominent beak.

Sedges flower in late May and June.

5
3
4
2

Common Watercress Cruciferae
Nasturtium officinale R. Br.

Watercress was considered a very nutritious vegetable even in the days of the ancient Greeks and Egyptians and not long ago it was grown in Europe in large quantities. Many gardeners specialized in the cultivation of Watercress, particularly in the vicinity of Paris and Erfurt. From there it was shipped either fresh or in a salt and vinegar solution primarily to England. The English were the chief customers, for English cooking used Watercress not only in salads but also served with meats and it was not just the fleshy leaves that were used in the kitchen — the seeds were added to mustards for flavour.

The cultivation of Watercress naturally required different methods to those commonly used in gardening. Since it is a typical waterside herb it is closely tied to water. For that reason it was grown in canals about 2—3 m wide where the level of the water was regulated by sluices and the average temperature of the water was kept at a mere 8° C.

Watercress grows particularly well in cold water and is often used in trout hatcheries where it keeps the growth of fibrous algae in check and enriches the water with oxygen.

Because of the vitamins it contains Watercress was also used as a medicinal plant. Even today it continues to be recommended for stimulating the appetite and for digestive disorders. Formerly, Watercress was also used to prevent scurvy where there was a lack of vitamin C at certain seasons.

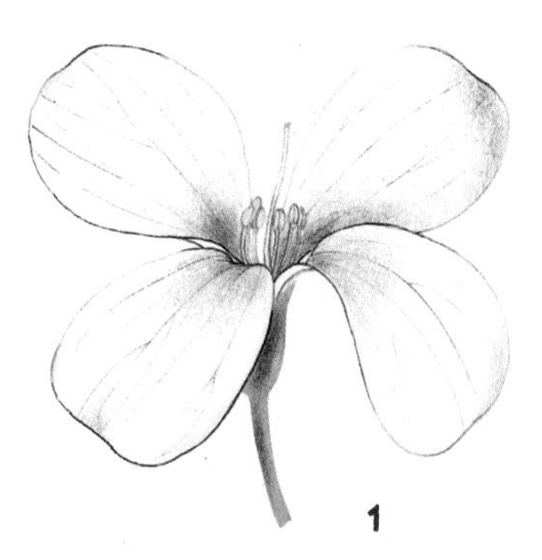

Watercress is a cosmopolitan plant distributed in Europe as far north as Denmark. It has a hollow, sharply grooved stem that is 20—90 cm high and often branched; it roots at the base (when it is flattened to the ground). The odd-pinnate leaves, with large terminal leaflet, are soft and succulent. The flowers are small and white with prominent yellow anthers (1). The fruit is a cylindrical, slightly sickle-shaped siliqua with 2 rows of seeds. The top parts contain glycosides and bitter principles, and eaten fresh are a source of vitamins A, C, and E.

The flowering period is from May to the beginning of July.

Brooklime
Veronica beccabunga L.

Scrophulariaceae

Brooklime is found by water-filled ditches, on stream banks, pond and lake shores and round springs. It does best in the vicinity of cool, oxygen-rich water. Like many shoreline plants it not only occurs as a terrestrial and submerged form but is otherwise also quite variable. Usually, however, these others are dwarf forms occurring in dry habitats and likewise influenced by ecological conditions.

Brooklime is most widely distributed in lowland and hilly districts; rarely is it found at elevations above 1,000 m (in the Alps above 2,000 m), even though it would find the otherwise cool mountain streams congenial.

Of the 15 species of the *Beccabunga* group of the genus *Veronica* Brooklime (*V. beccabunga*) and Water Speedwell (*V. anagalis-aquatica* L.) (1) are the most widespread. Brooklime is distributed in practically all of Europe, western and northern Asia and north Africa, Water Speedwell has a world-wide distribution thanks to man.

The fruits of plants of the *Beccabunga* group are often attacked by weevils and so deformed that they resemble blueberries, both in size and colour.

1

The derivation of the generic name *Veronica* is very complex; however, the specific name *beccabunga* is an example of a latinized German common name — 'Bachbungen' — which traces its origin to a medieval German dialect.

Brooklime is 10—60-cm-high perennial herb. The leaves, either all or at least the bottom ones, have stalks and blunt rounded tips. The flowers, 5—8 mm in diameter, are arranged in racemes that are only 1 to 3 times longer than the subtending bracts. A raceme consists of as many as 25 flowers. The fruit is a broadly heart-shaped inflated capsule about 3—4 mm in diameter.

The flowering period is from May until late summer.

Yellow Flag
Iris pseudacorus L.

Iridaceae

Yellow Flag, of the iris family, is not a very common plant. Despite the fact that it multiplies readily by vegetative means it is rapidly disappearing from its original habitats and in many countries is therefore on the list of protected species. Partly responsible for its decline in number in the wild were the herb healers of old who collected and prescribed its 'root' (rhizome) as a remedy to halt bleeding, e.g. in the case of haemorrhoids (according to Mattiolli's Herbal). When Sweet Flag (*Acorus calamus*), of the arum family, was discovered in Europe *Radix Acori palustris* was the name given to the root of Yellow Flag and *pseudacorus* (or false acorus) the designation applied to the plant itself. This was doubtless due to the striking resemblance between the leaves of the two plants as well as several other common characteristics. The rhizome of Yellow Flag was also put up in wine and used as a remedy for jaundice.

The overwintering leaves serve as excellent examples demonstrating wilting caused by cold. The leaves, which are normally rigid, become flaccid and droop at low temperatures. This is caused primarily by physiological dryness, for the green leaves transpire even in winter but the underground parts cannot absorb water in the frozen soil. A similar phenomenon may be observed in evergreen plants of a woody nature and in winter grain crops.

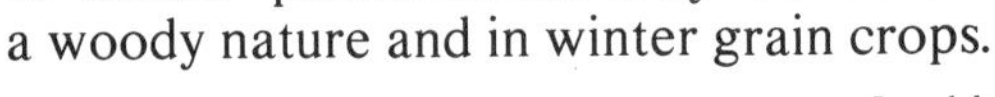

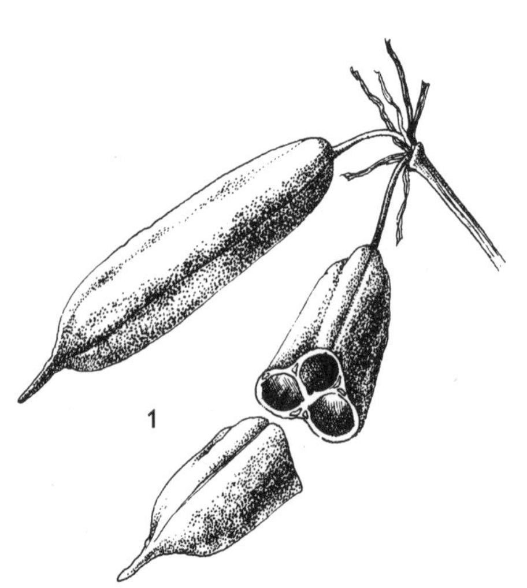

In old Greek *íris* was the word for rainbow. The flowers of many irises, particularly the modern garden hybrids, are noted for their striking coloration. Yellow Flag is a humble plant in comparison.

It is a 50—100-cm-high perennial with a thick, branched rhizome. The erect stem, slightly flattened and branching at the top, is terminated by long-stalked flowers growing from the axils of green bracts. The leaves are sword-shaped and usually longer than the flowering stem. The fruit is a large, cylindrically three-sided capsule with pointed tip (1).

Yellow Flag grows scattered at the edges of ponds and in waterside thickets as well as alder groves in practically all of Europe, excepting the arctic regions; its distribution extends to western Siberia.

The flowering period is in June.

Sweet Flag
Acorus calamus L.

Araceae

Nowadays we cannot imagine a pond or lake without the fragrant Sweet Flag, but this was not always so. In the days when most of Europe's ponds were being established (14th to 16th century) there was no trace of this plant on the Continent. At that time it grew somewhere in eastern India and China but today no one knows for sure. It was not until the late 16th century that the first rhizomes were brought from Constantinople to Vienna when the plant soon spread to fresh waters throughout Europe.

When not in flower Sweet Flag readily escapes notice, particularly amidst reedmace, irises and bur-reeds — but not if a person has a keen sense of smell. The aroma of the leaf of Sweet Flag rubbed between the fingers is unique and inimitable, the same as that of the thick rhizomes. It is the latter that made Sweet Flag important as a medicinal plant and flavouring agent; in ancient Hindu, Arabian, Greek and Roman medicine and confectionery the root of Sweet Flag was a very useful drug recommended as an ingredient of herbal tea mixtures stimulating the appetite. It is also recommended in the form of baths to calm the nerves.

Sweet Flags introduced to Europe (and America) produce flowers occasionally but the seeds do not mature. All existing masses of Sweet Flag are thus clones, produced by vegetative means from a single individual.

Sweet Flag is a member of the arum family, distinguished by the characteristic structure of the inflorescence; best known of the lot are the Calla Lilies, used in bridal bouquets, and Anthuriums. The minute flowers (1) are clustered on a fleshy spike, called a spadix, usually enclosed in a strikingly coloured spathe (white in Callas and red, as a rule, in Anthuriums). Sweet Flag is an exception. Its spathe is green, looks very much like a leaf and is a continuation of the stem. The actual inflorescence — the spike — though terminal, thus appears to be located at the side of the stem (2). The leaves of Sweet Flag are shining and strap-shaped with the margin sometimes conspicuously 'finely wrinkled'.

The rhizomes contain up to 4 per cent essential oils and bitter principles, with tannins being produced during drying. The characteristic aroma is caused by aldehydes.

The flowering period is in June and July.

2

Reed Sweet-grass
Glyceria maxima (Hartman) Holmberg

Gramineae

Sweet-grass is the third largest and most common component of high-stemmed waterside reed-beds, after reed and reedmace. Like them it, too, was used by man and played an important role in his life. In early days the grains of Floating Sweet-grass (*G. fluitans* (L.) R. Br.) were gathered in times of need and ground into a fine meal ('manna'); the leaves of Reed Sweet-grass were used for thatched roofing.

Reed Sweet-grass is a strongly competitive plant that forms monotonous spreading masses in the shallow backwaters of ponds and reservoirs. It tolerates low temperatures and thus has a lengthy growing period. Because of its great powers of propagation by vegetative means it also tolerates occasional cutting or damage and nibbling by muskrats. In deep and flowing water, e.g. in ditches supplying water to ponds, it forms submerged leaves carried by the current and does not flower.

A more welcome species in pisciculture is Floating Sweet-grass which serves as an indicator of good soil conditions and provides fish with shelter as well as suitable places for spawning.

Sweet-grasses provided good, sweet fodder for cattle and horses and that may also be the reason they were given the name *Glyceria* (the Greek word *glykerós* means sweet); another theory links the name with the grains of Floating Sweet-grass which contain about 40 per cent digestible sugars.

Reed Sweet-grass (1a) is a perennial grass often up to 2.5 m high with a long, creeping, stoloniferous rhizome located at a shallow depth. The rigid stems are conspicuously smooth and shining. The inflorescence is a relatively large, loose panicle. The spikelets (1b) are set at an angle and composed of numerous flowers coloured yellowish-green or tinted violet. Floating Sweet-grass is likewise a perennial. The floating leaves may be 50—200 cm long. The flowering stems are only 30—100 cm long and terminated by a one-sided panicle (2) with branches practically horizontal in full flower. After the grains mature, however, the branches right themselves and are once again pressed to the main axis.

Sweet-grasses flower from May to August.

2
1 a

Reed Canary-grass
Phalaris arundinacea L.

Gramineae

Reed Canary-grass is often mistaken for Reed, particularly when it is not in flower, for it slightly resembles that plant both in the colour and roughness of the leaves. It often grows by the waterside, in ditches and in wet meadows. It does not mind irregular fluctuations in the water table and even appears to find this congenial. It is very common on river banks where it forms typical communities of meandering and calm water set apart from the main stream and sometimes incipient communities in places that are regularly flooded.

Practice has proved Canary-grass may be good fodder for horses and cattle. It was said to make a horse's hide gleam and butter made from the milk of cows fed on Canary-grass had an unusually fine flavour. That was the reason Canary-grass was also cultivated in some meadows that were difficult to drain.

Reed Canary-grass is distributed throughout practically all of Europe, western, northern and eastern Asia and North America. It is thus a circumpolar species; its occurrence in south Africa is sometimes considered to be secondary.

Decorative forms are occasionally grown in cultivation: the variety 'Picta' has variegated leaves, the variety 'Elegans' has leaves striped green and white.

Canary-grass is a robust perennial grass with creeping rhizome. The stems are terminated by long, narrow, lobed panicles coloured whitish-green tinged with red before flowering and later straw-yellow, the same as the whole plant as it dries (1). They do not disintegrate, however, but remain on the stem until the following season, thus giving stands of this grass their characteristic appearance.

The spikelets that make up the inflorescence are one-flowered (2) and hermaphroditic, with remnants of two atrophied flowers.

The flowering period is in June and July.

2

Soft Rush
Juncus effusus L.

Rushes are faithful companions of water. The cosmopolitan Soft Rush not only has a continuous distribution in Europe and on the Atlantic coast of North America, but also occurs on the Pacific coast, in South America and Africa as well as in a large part of south-east Asia.

Rushes grow not only by the waterside but also in woodlands and meadows. Their presence is always an indication of a stagnant water table or damp, clay soils. In cultivated meadows they are an unwelcome weed. They are likewise an unwelcome sight to fishermen on the banks and in the shallow waters of ponds for they have large root systems, spread rapidly by vegetative means, and thus contribute to the filling-in of shallow bodies of water.

The stems of larger rushes, such as *J. effusus*, are usually rigid and erect; on the European coast of the Atlantic one will find bizarre stands of Soft Rush with stems irregularly twisted and coiling in spirals.

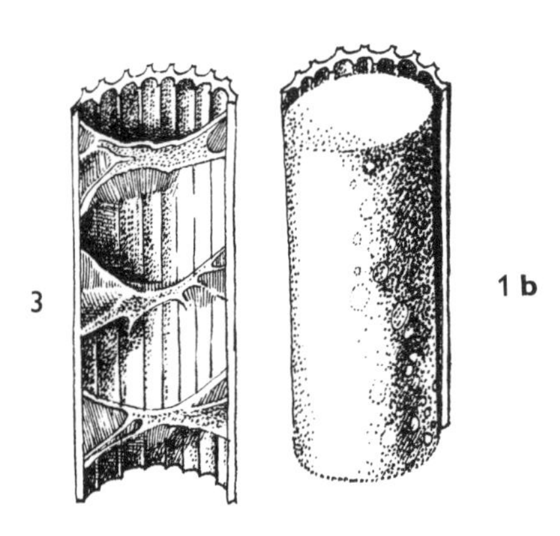

Rushes are annual or perennial herbs with firm, tough stems formerly used for binding and weaving, hence also the generic name *Juncus*, derived from the Latin word *jungere* meaning to bind.

Juncus effusus (1a) is a densely tufted perennial herb, sometimes only 30 cm high and at other times more than 1 m high and usually coloured a fresh green. The stem is rounded and circular in cross section; it is without leaves but has dark-red sheaths at the base and a continuous, white cottony pith inside (1b). The leaves, when present, are similar. The inflorescence is loosely spreading and seemingly lateral (because the subtending bract continues on far above the stem).

The inflorescence of the closely related *J. conglomeratus* L., found only in Europe, is ball-like (2). Both the aforesaid rushes are usually accompanied by another, whose greyish-green colouring makes it conspicuous from afar — namely Hard Rush (*J. inflexus* L.). This, however, has an interrupted pith as seen on the longitudinal section (3).

Rushes flower from late May onward, sometimes until October.

1 a
2

Great Spearwort
Ranunculus lingua L.

Great Spearwort is one of the largest of the many members belonging to the genus, which numbers some 800 species, and also has the biggest flowers. It grows scattered in central and western Europe on the boggy shores of ponds and lakes, in water-filled ditches and in pools. It appreciates occasional flooding. Like Amphibious Bistort it occurs in two forms — terrestrial and aquatic (submerged) and is generally found in spring or else not until autumn at depths of about 30—60 cm.

The buttercup family is a systematic group characterized by, amongst other things, having flowers with a great many stamens. The pollen grains in a single anther are at the same stage of maturation but those of various anthers exhibit marked differences — in some the development of pollen is in its initial stage, in others the pollen grains are fully developed. This naturally plays an important role in pollination, especially in view of the long time the flower remains open.

Many buttercup species contain antibiotic phytoncids which inhibit bacterial multiplication in a fresh juice. However, they are slightly toxic and may cause blisters in persons with sensitive skin.

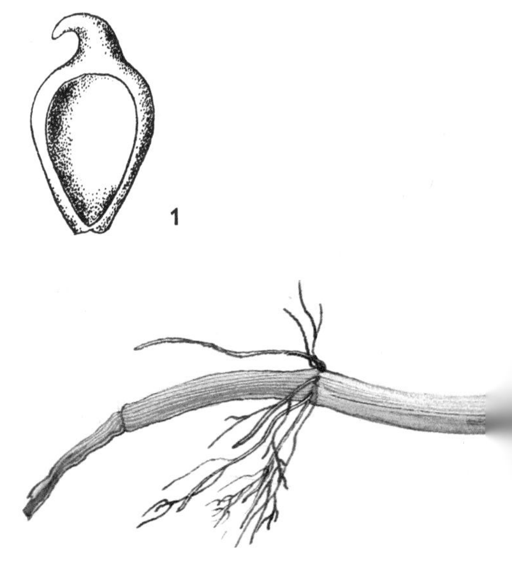

Ranunculus lingua is a robust herb, growing sometimes as high as 150 cm with a jointed, hollow rhizome and numerous underground shoots. The stem is not greatly branched, as a rule, and is covered with rigid, lanceolate, entire leaves. This shape is not common in the buttercup family, which generally has palmate or otherwise divided leaves. It is also reflected in the name of the species, for the Latin word *lingua* means tongue (after the 'tongue-shaped' leaves).

The flowering period is from June to August. The fruit is an achene with hooked beak (1).

Arrowhead
Sagittaria sagittifolia L.

Alismataceae

The genus *Sagittaria* includes some 30 species, most of them American. In Africa and Australia they are absent in the wild. Some species, such as *S. latifolia* and *S. isoetiformis*, however, are found on all continents where they are popularly grown in aquariums. The unusual shape of the leaves has also been responsible for their being grown in gardens, where they are even cultivated as a double form.

In Eurasia the genus is represented by *S. sagittifolia* L. which is distributed from the western coast of Europe to the islands of eastern Asia, usually occurring only in lowland districts, however. It is rarely found at elevations of more than 550 m. It grows in still and slow-flowing water, in ponds and lakes. The places where it occurs are all relatively muddy so that its presence signals the silting-up of a reservoir. Stands of Arrowhead, however, are popular with fish — as a source of food and place to spawn.

Arrowheads generally multiply by vegetative means. During the summer they form starchy overwintering buds on the runners that are up to half a metre long. The parent plant dies, but in spring these buds give rise to a whole new colony of arrowheads.

The species *S. arifolia* Sm. of North America is very similar to the European Arrowhead in many ways.

The scientific as well as common names of these plants are derived from the typical shape of the leaves which resemble an arrowhead — the Latin word *sagitta* means arrow. That, however, is the shape of leaves that rise above the surface. In deep and more swiftly-flowing water the plant bears only long, narrowly-linear leaves (1). That is also what the first young leaves look like. In calm water these are followed by leaves of the 'second generation' with relatively small, long-stalked blade that floats on the water when it reaches the surface (2); only later are the conspicuously arrow-like leaves (3) formed by older plants in shallow water.

The plant parts growing in mud and underwater are provided with ample oxygen through aerating tissues with large intercellular spaces (4). The flowers (5) of arrowheads are of two kinds: long-stalked male flowers and short-stalked female flowers. The fruit is an achene. Arrowhead rarely bears flowers (in shallow water only) from June to August.

4
1
2
3

Water Horsetail

Equisetaceae

Equisetum fluviatile L.

Horsetails are evolutionary relics amongst existing plants. In the figurative sense they are witnesses of the early geological periods of the Earth's development. The ancestors of today's herbaceous horsetails contributed to the formation of coal deposits which is, at present, a chief source of energy.

Of the existing horsetails the ones most closely tied to shoreline and shallow water in Europe are Water Horsetail and Marsh Horsetail (*E. palustre* L.). The first often grows on peaty substrates (particularly at lower elevations in foothills) in plant communities at the margin of still water where, in the initial stages of reed bed formation, it may be the prevailing species.

Marsh Horsetail often occurs in marsh meadows and damp pasturelands in clay soil, mainly at lower elevations; only occasionally is it found at higher altitudes, even though in the Alps it is encountered at elevations od 2,450 m. Both species are circumpolar in the northern hemisphere.

Because of their rapid spread by means of underground runners — jointed creeping rhizomes that penetrate to depths of 1 m — horsetails help colonize littoral zones; they are very vigorous and tolerate frequent cutting. A single square metre of cut marshland, for example, was found to contain 1,300 stems of Water Horsetail! Because of the great amount of silicic acid in its cell membranes it is not suited for fodder.

Horsetails were named by the ancient Greeks which is how they come by their latinized name *Equisetum* — the Latin word *equus* means horse and the word *saeta* means bristle. They are perennial herbs with long-jointed rhizomes and erect stems branching in whorls. The stems are usually grooved and also prominently jointed, the base of each joint enclosed by a scale-like leaf (sheath). In horsetails that are covered by water for a time they often root at these joints (1).

Some horsetails have two types of stems: one in spring and the other in summer. Water Horsetail and Marsh Horsetail have only a single green stem terminated by a spore-bearing cone (2). The spore-bearing cones appear at the tips of the stems in early summer.

Marsh Horsetail has slender, conspicuously grooved stems and outspread (not appressed) sheaths with 6—10 teeth (3); Water Horsetail has thicker, rounded stems and appressed sheaths with 15—30 teeth (4).

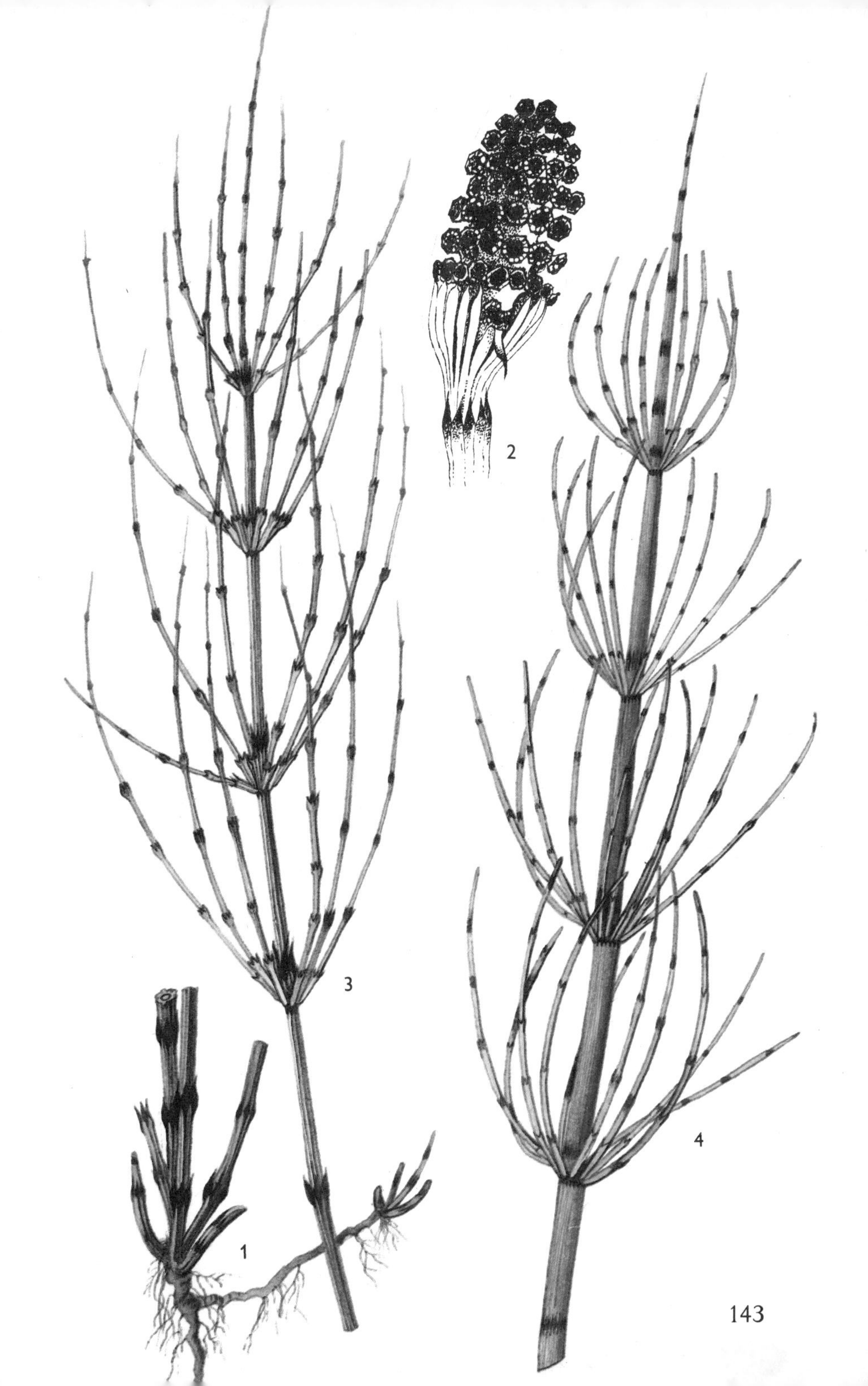
2
3
1
4

Common Blinks Portulacaceae
Montia fontana L.

Gardeners making great efforts to obtain the dry-loving Bitter-root from North America for their rock garden may have no idea that they can find a related species from the same family growing by Europe's streams, namely Common Blinks. And no wonder, for montias are very inconspicuous, small plants greatly resembling certain chick-weeds. Because they are plants of fresh, clean water they are rapidly disappearing from their habitats due to man's incursions.

Montias are most common by mountain streams and springs and on the flooded sands of the upper reaches of rivers and brooks. The leaves are slightly fleshy and masses of these plants form a thick, deep-green carpet providing shelter for young trout fingerlings. At the same time such spreads enrich the water with oxygen.

Common Blinks has a continuous distribution throughout north-western Europe and grows intermittently on the eastern coast of North America, the Greenland coast, Alaska, and the islands of eastern Asia (Kuril, Sakhalin, Japan).

Montia fontana is a relatively variable species — some strains are annual, others biennial or even perennial — and it is also marked by seasonal dimorphism: besides the typical spring populations there are also autumn populations that differ in appearance.

The genus *Montia* was named after the 18th-century professor G. Monti who lived and worked in Bologna, Italy.

It is 10—20 cm high with opposite, entire leaves and small white flowers in terminal or lateral inflorescences. The flowers consist of a 2-merous calyx, two large and three small petals, and three stamens (1). Autogamy (self-fertilization) was observed in flowers in cloudy weather without their even opening. The mature seeds are ejected from the splitting capsules to a height of half a metre and quite some distance away. The flowering period is in June and July, or in September and October.

Common Blinks often grows by stream-sides together with Bog Stitchwort (*Stellaria alsine* Grimm), Floating Sweet-grass (*Glyceria fluitans* (L.) R.Br.) and Brooklime (*Veronica beccabunga* L.). It is often mistaken for Bog Stitchwort (2) but the latter has pointed leaves and regular 5-merous flowers in inflorescences whereas *Montia* has blunt leaves and solitary flowers.

1
2

Branched Bur-reed
Sparganium erectum L.

Sparganiaceae

The bur-reed (Sparganiaceae) family is a very old one; fossil species were present on the Earth as far back as the Mesozoic (Cretaceous) period on the territory of present-day Greenland.

Today the family is represented by only a single genus – *Sparganium* – which includes some 20 species distributed throughout the north temperate zone and one species in New Zealand. Of the several native to Europe the commonest is the Branched Bur-reed (*S. erectum*, syn. *S. ramosum* Huds.). Besides many other unusual features in the plant's shape and anatomical structure, noteworthy is the structure of the leaf blade. The tissues where photosynthesis takes place are distributed round the whole periphery of the blade (3-angled in cross-section) and thus the leaf's capacity to carry out photosynthesis is extremely great. The seasonal biomass production, however, is relatively low compared to that of other emersed herbs and the growth period is quite short. Masses of this bur-reed, spreading by vegetative means, often occupy the open spaces of shoreline waters and are sensitive to the competition of other taller plants, chiefly reeds. This probably goes hand in hand with the great light requirements of the genus *Sparganium*. Thick masses of bur-reed, however, contributed to the rapid coverage of bodies of water with vegetation and their tolerance of low temperatures and freezing-over of the water thus makes up for their aforementioned handicap in competition with other plants. The leaves, sometimes more than 2 m long, were formerly used to make baskets.

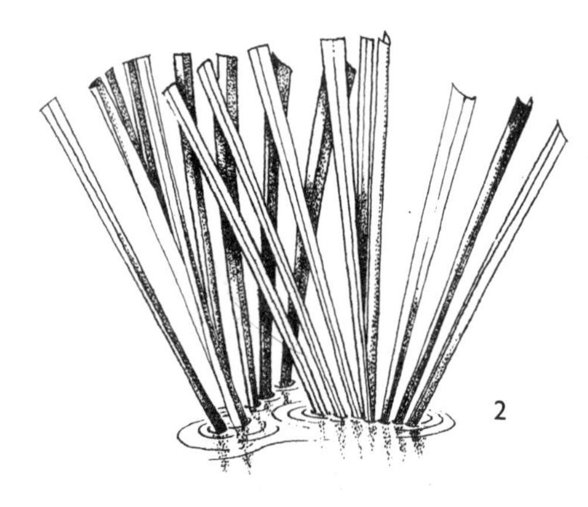

Sparganium erectum is a perennial herb 30—200 cm high with creeping, stoloniferous rhizome and linear to sword-shaped leaves three-angled in cross-section (1). The stem is usually shorter than the leaves, 50—150 cm high, and branched. The leaves of non-flowering (sterile) bur-reeds are sometimes arranged in the shape of a fan (2); in deeper or flowing water the narrow leaves float on the surface. The dioecious flowers are arranged in globose heads. The ones at the bottom are larger and composed of female flowers, those farther up are smaller and composed of

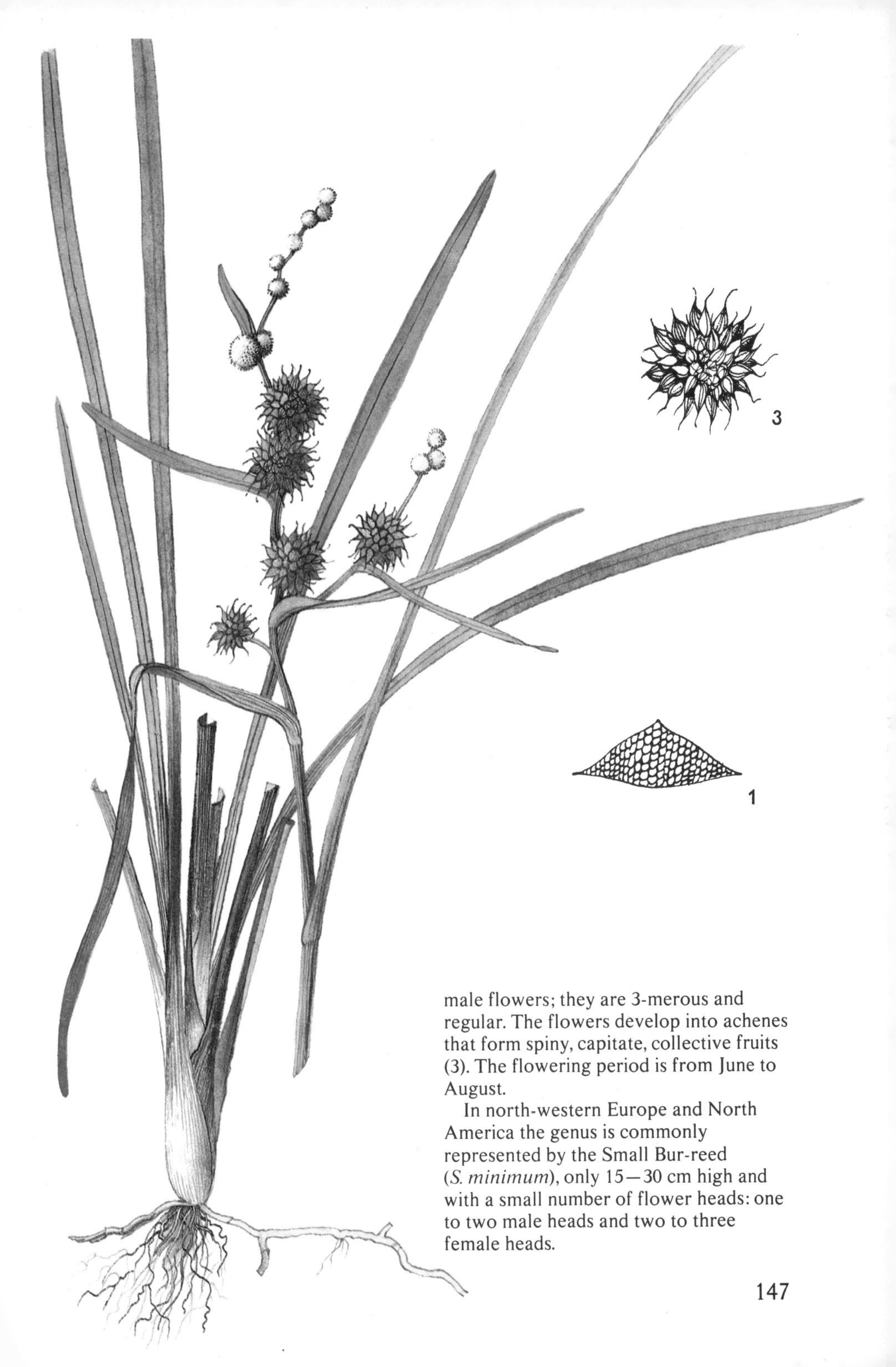

male flowers; they are 3-merous and regular. The flowers develop into achenes that form spiny, capitate, collective fruits (3). The flowering period is from June to August.

In north-western Europe and North America the genus is commonly represented by the Small Bur-reed (*S. minimum*), only 15—30 cm high and with a small number of flower heads: one to two male heads and two to three female heads.

Flowering Rush
Butomus umbellatus L.

Butomaceae

Flowering Rushes are tied to water and waterside bogs. Their only representative — *Butomus umbellatus* — is distributed throughout practically all Europe, except the more northerly parts of Scotland and Scandinavia; it is also absent in high mountains. Its continuous range, however, extends far into central Asia.

Flowering Rush is also tied to rather warm lowland and hill districts. It grows in pools, ponds and lakes as well as by rivers, in still water and shoreline reed beds — best of all in places with muddy bottoms. It is readily cultivated and a popular plant of garden pools, but may spread and completely dominate them after a time — its decorative flowers, however, are worth it. The leaves are narrow, strap-shaped and very rigid and were formerly used in weaving mats and baskets, the same as bur-reed and reedmace.

In former times the slightly bitter rhizome was used in folk medicine under the name *Radix Junci floridi* given it by our ancestors because its leaves and the places where it grew reminded them of rush *(Juncus)* — one with brightly coloured flowers, however.

Flowering Rush is a perennial herb with a thick rhizome which bears narrowly-linear leaves, sheath-like and three-angled at the base and gradually expanding into a flat blade sometimes up to 2 m long.

The flowers are arranged in an umbel-like inflorescence borne at the tip of a rounded scape. They are regular, hermaphroditic, and 3-merous (1). After pollination and fertilization the stigmas develop into follicles (2) containing light, floating seeds. Though these have good powers of germination even after a lengthy period out of water, Flowering Rush usually multiplies by vegetative means (root buds).

The flowering period is from June to August.

2

Fine-leaved Water Dropwort Umbelliferae
Oenanthe aquatica (L.) Poir.

Water dropworts are usually large, robust plants growing singly, in large groups or even in continuous masses on the edges of ponds and pools and in shoreline reed communities. They are generally winter herbs whose seedlings appear at the beginning of autumn, forming striking and decorative ground rosettes, particularly on the exposed bottoms of drained ponds. When the latter are filled again they can exist as submerged forms — sometimes remaining thus for the whole of the next growing season. Generally, however, they form stouter, hollow stems the second year. Water dropworts can also grow on the exposed bottoms of ponds that remain drained for a long time but then they are always of smaller, weaker habit.

Fine-leaved Water Dropwort is a Eurasian species with Europe its centre of distribution (excepting the arctic regions). It was introduced to North America. Formerly the 'seeds' (i.e. the fruits) were used in folk and veterinary medicine to cleanse wounds and in compresses.

Fine-leaved Water Dropwort is an annual or biennial herb with hollow, grooved stem terminating abruptly at the base and with spreading branches; it often roots at the nodes underwater (1). The compound, odd-pinnate submerged leaves at the base (2) differ in size and in the width of the segments from those higher up. The

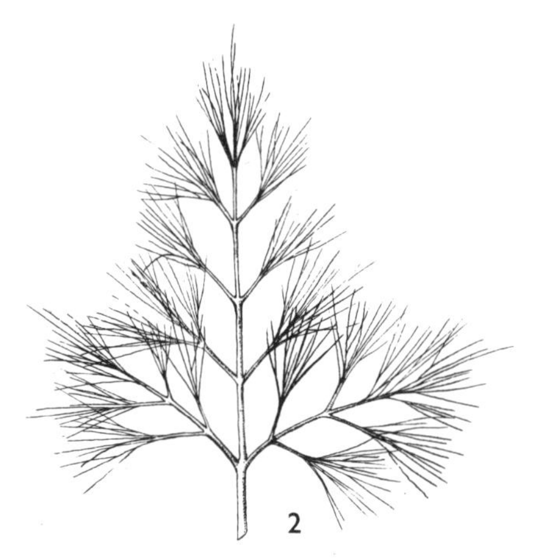

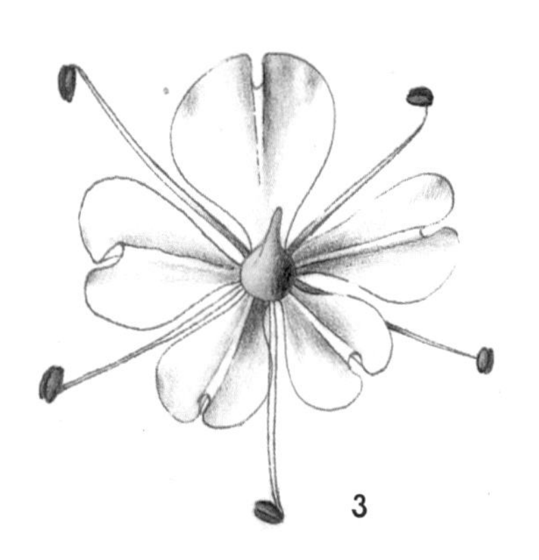

flowers (3) are hermaphroditic, only faintly radiate and arranged in dense umbels. The fruit is a pendent double achene (4) with characteristic arrangement of the tissues in cross-section (5).

The flowering period is from June to August.

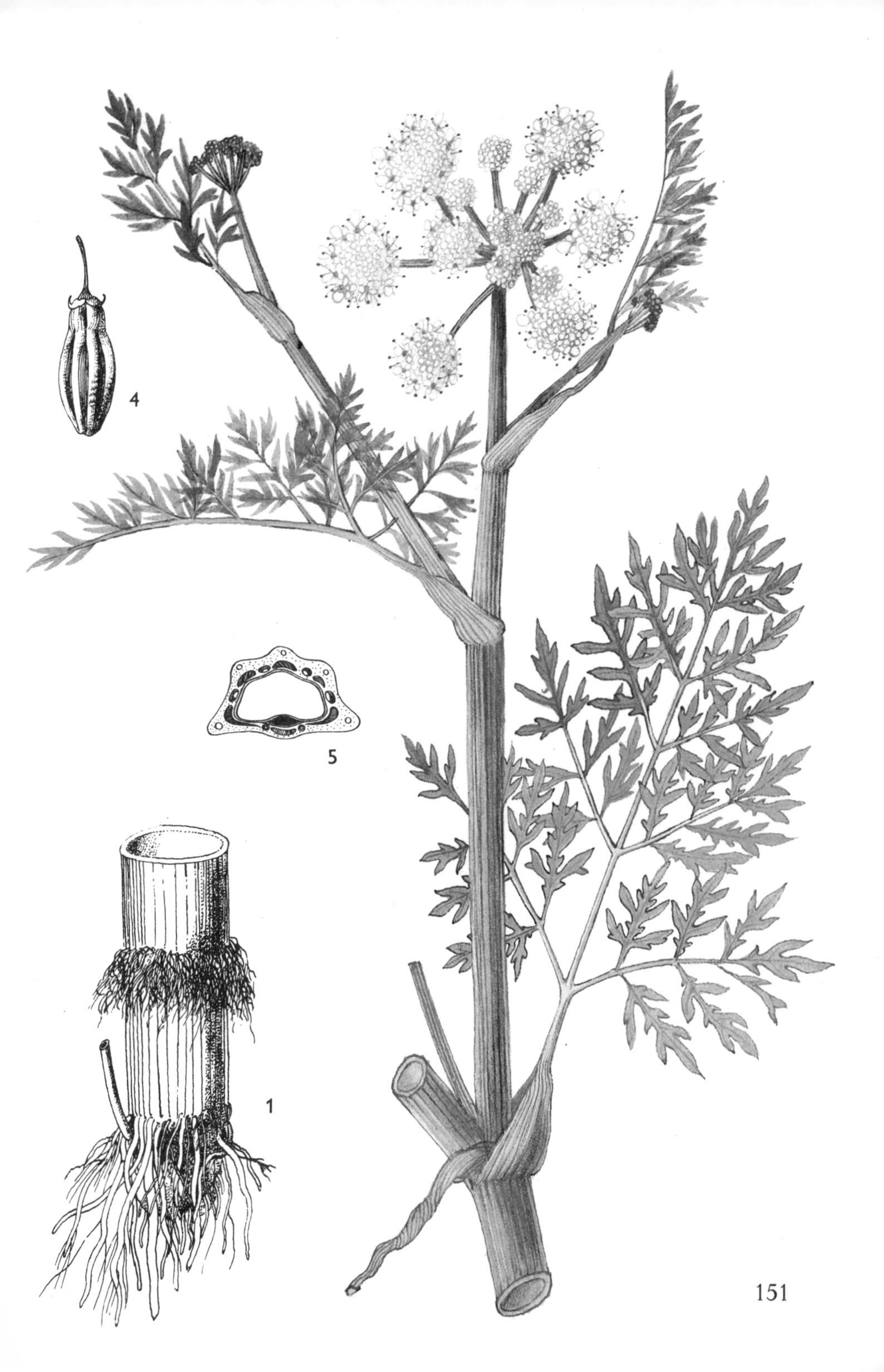
4
5
1

Common Water-plantain
Alisma plantago-aquatica L.

Alismataceae

The life of plants of the Alismataceae family is inseparably linked with water. This is a cosmopolitan family found on all continents, excepting the Antarctic. Common Water-plantain also has a widespread distribution — it is found on all the continents of the northern hemisphere with the exception of the most rugged arctic regions. Fossil finds of the genus *Alisma* dating from the Tertiary period, however, have been found in the Arctic.

For a plant that is sometimes quite large — it may reach a height of more than 1 m — it has a relatively brief growth period: the flowers fade and seeds are produced two months after growth starts. However, it also multiplies asexually by means of bulbils. Their good powers of reproduction enable water-plantains to be pioneer bog and shore plants that are among the first to appear in newly established or renewed bodies of water — they were found growing together with reedmace in a puddle in the middle of a little-used road. They do best in shoreline marshes and in calm, shallow water. They may also occur as submerged forms at greater depths. Their root systems are large and firm.

In old pharmacopoeias and herbals Water-plantain was listed as a medicinal plant and prescribed in the form of packs to relieve headache and internally as a remedy for rabies. The fresh roots and top parts are slightly poisonous.

Water-plantain is a perennial herb whose size depends on many external factors: sometimes it may be only 10 cm high, at other times it may reach a height of 1 m. It has a tuberous, thickened rhizome from which rises a huge clump of long-stalked leaves. The submerged leaves may be only linear with transitional forms to floating leaves, the same as in the related arrowhead (*Sagittaria*).

The inflorescence is usually large and panicle-like; the lateral branches are arranged in whorls consisting of three or more branches. The flowers are 3-merous (1) and hermaphroditic. They are produced from June to September. The fruits (achenes) mature only on plants with flowers above the water.

Skullcap
Scutellaria galericulata L.

Labiatae

Damp meadows, ditches, stream-, river- and pond-banks, edges of bogs and occasionally also exposed bottoms of ponds are all the sort of places where Skullcap will be found. Thence it often makes its way to damp alder groves and along large rivers even to cities; although the river-banks in cities are bound with stone and concrete there is always a crack or crevice somewhere! From lowland districts, where it is common, its distribution extends alongside water courses (carried by migratory water birds) to mountains; in alpine valleys it occurs even above the 1,000-m mark.

Skullcaps are an isolated group within the mint (Labiatae) family. The approximately 180 to 200 species that make up the genus *Scutellaria* are distributed in the temperate and tropical regions of the whole world. The illustrated species is circumpolar and widespread throughout practically the entire northern hemisphere, its continuous range extends to 69° latitude North in Scandinavia and southward to the northern Balkans and northern Italy. It is a relatively variable species often occurring in various forms adapted to various environments; for instance plants growing in shady situations are usually very tall (up to 75 cm high) with long leaves. Like most plants that grow beside water there are also submerged aquatic ecotypes adapted to flowing water. Formerly *Scutellaria galericulata* was used in folk medicine as a remedy for malaria.

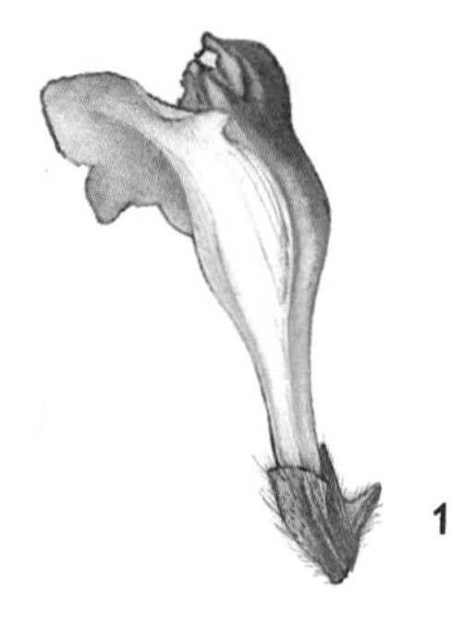

Skullcap derives its generic name *Scutellaria* from the peculiar shield-like growth or scutellum on the lip of the calyx. It is a stoloniferous perennial herb, usually 20—40 cm high, with indumentum that exhibits marked variability; sometimes it is glabrous, at other times hairy. The stem is rarely branched, the leaves are opposite and short-stalked.

The flowers (1) generally grow in twos (though there may be anywhere from one to four) in one-sided verticillasters from the axils of the leaves. The corolla tube is usually very long and curved so that the flowers can be pollinated only by insects with a long proboscis (bumblebees, butterflies).

The flowering period is from June to September.

Water Mint
Mentha aquatica L.

Labiatae

Mints, along with Sweet Flag, are the most fragrant 'water' plants. They often grow in masses that spread by vegetative means in ditches, marshy places, willow thickets and reed beds by flowing as well as still water. Water Mint is distributed throughout practically all Europe, north-western Asia and south Africa; on other continents it is a naturalized, not indigenous, species.

Mints pose a problem for botanists for they are extraordinarily alike and often interbreed. The hybrids are also fertile and cross-breed further themselves. It is therefore not surprising that authorities differ in their opinion as to the number of existing species — some say there are barely 15 whereas others assert there are as many as 600.

Mints are age-old medicinal plants. Most often prescribed in pharmacopoeias is the drug (leaves or top parts) of Peppermint (*M.* × *piperita* L.), which is probably a hybrid of Water Mint and Spearmint (*M. spicata* L.). The drug from Peppermint alleviates spasms in digestive disorders, prevents flatulence and stimulates the flow of digestive secretions; it also has an antiseptic and anti-inflammatory effect. It is used in the form of an infusion, externally as a bath preparation, and also as a pleasant, mint-flavoured tea. Long-term usage, however, may be harmful.

Water Mint is a softly-hairy perennial herb with numerous underground shoots. The stems are ascending, little-branched, and 25—100 cm high. The fragrant leaves are stalked and ovate to lanceolate in outline. The flowers grow in semi-globose verticillasters from the axils of the uppermost leaves and at the tips of the stems. The corollas are usually bright violet, pink or white; they consist of a short tube with four-lobed edge; the four equally-long stamens and the stigma are exserted, i.e. they protrude from the flower (1). The plant tissues contain mainly essential oils (menthol, menthylester, menton, etc.).

The flowering period is from June to October.

1

Bulrush

Cyperaceae

Schoenoplectus lacustris (L.) Palla

The Bulrush gives lakes and ponds a slightly exotic look. Where it is part of shoreline reed-beds, i.e. stands of reeds and reedmace, it usually escapes notice. Unlike the latter, however, it tolerates water of varying depths and thus it is not surprising that in deeper water it is the only tall-stemmed emersed marsh grass. In such bodies of water one can see occasional small bulrush colonies that appear to be floating freely on the surface. This, of course, is merely an illusion for the leafless stems of Bulrush often measure about 3 m in length and tufts of these stems are firmly anchored in muddy as well as in sandy or stony bottoms. The dark-green leafless stems then jut above the water like strong wires.

Bulrush is indigenous to and distributed throughout practically all Eurasia. However, it also grows on other continents, except in arctic regions. Bulrush does not grow only in lakes and ponds as one would think from the Latin name *lacustris,* meaning lake. Pure stands may be found also on the banks of streams, ditches and rivers. In such places the otherwise rigid stems (and also possibly developed narrow leaves) are bent by and appear to be flowing with the current.

Bulrush does not tolerate lengthy dry periods and becomes atrophied; however, when covered by water again its strength and vigour will return.

Bulrush is a perennial herb 80 cm to 3 m high with a thick, jointed, creeping rhizome and robust, rounded stems enclosed by violet-brown sheaths at the base. There are always several such stems rising from the large root system and they thus form typical, tufted colonies (1). The inflorescence is a terminal, branched anthella with a channelled and pointed subtending bract longer than and extending beyond the inflorescence. The flowers are hermaphroditic with ciliate bracts and perianth bristles, three stamens and three stigmas (2).

The flowering period is from June, sometimes until October.

Continuously distributed in North America and only occasionally in Europe is the species *Sch. americanus* (Pers.) Volk. with stems three-angled beneath the inflorescence, leaf blades up to 20 cm long and stalkless spikelets.

1

Great Reedmace
Typha latifolia L.

Typhaceae

Reedmaces are found practically wherever water collects at least for a while, sometimes even in a puddle by the roadside. Probably the commonest and most widespread is the Great Reedmace found throughout the world — hence circumpolar — and distributed throughout the whole of the northern hemisphere as well. Lesser Reedmace (*T. angustifolia* L.) is also a circumpolar species with eastern North America, Europe and north-western Asia the focal points of its distribution.

Reedmaces are among the Earth's large producers of biomass. They generally form large, pure masses at the edge of still and slow-flowing water and may be permanently covered by water up to half their length. They spread rapidly by means of underground shoots and thus quickly colonize the shallow sections of bodies of water. Fresh reedmace was formerly used as supplementary fodder for cattle; the rhizomes were fed to pigs. In the present century muskrats have become a grave natural enemy for they are able to undercut even large spreading masses within a very short time.

The leaves of reedmace, cut before flowering and dried, were formerly used in making matting, hats, tote-bags, handbags, etc. Particularly noted for this type of product is Czechoslovakia, where the tradition continues to this day.

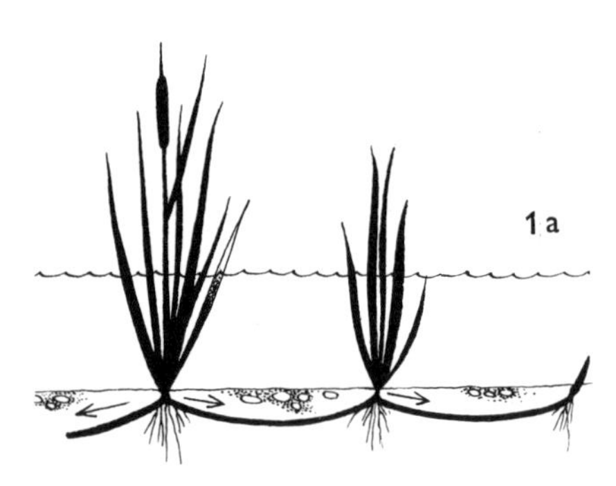

Reedmaces are perennial herbs, often of robust habit, which grow up to 3 m high. In muddy bottoms they spread by means of thick but soft rhizomes (1a) extending in all directions from the parent plant (1b). A stand of reedmace may even triple in size during a single growing season by this means.

The inflorescence is composed of two spadices: a female spadix below and a male spadix above. In the Great Reedmace the two are contiguous (2), whereas in the Lesser Reedmace they are 1—9 cm apart (3). The male inflorescence disintegrates soon after the flowers are spent and disappears, the female flowers develop into achenes with long hairs.

The greatly reduced flowers of reedmace — unisexual, without a perianth, and limited to only 3 (1—7) stamens or a single pistil — are borne in July and August.

1b
2
3

Great Water Dock
Rumex hydrolapathum Huds.

Polygonaceae

Dock was purportedly called 'rumex' by the ancient Romans and that is the name by which it continues to be known in scientific nomenclature. Docks grow in various environments but mostly in damp places. Great Water Dock is one of the largest and most striking. It grows on the banks of streams, ponds and rivers, in marshes, as well as in shallow ditches throughout western, central and eastern Europe. In the southern and northern parts it occurs only locally or not at all (it is absent, for instance, in northern Scotland), but, like many aquatic plants, it may be introduced even into these places.

In living organisms the number of chromosomes occurring in the nucleus of somatic cells is generally double the number normally occurring in the mature sex (male, female) cell. The former is called the diploid number and is designated by the symbol 2n; the latter is called the haploid number. In the first half of the present century, however, it was discovered that the cells in a great many of the large genera of the temperate zone were polyploid, i.e. the number of chromosomes in the somatic cells was three or more times the haploid number. The genus *Rumex* is a graphic example. Red-veined Dock (*R. sanguineus* L.) has a 2n of 20, Broad Dock (*R. obtusifolius* L.) is tetraploid and has a 2n of 40, and the illustrated Great Water Dock has 200 chromosomes in the nucleus of the somatic cells! These differences are naturally reflected in the external appearance and shape of the individual organs of such plants.

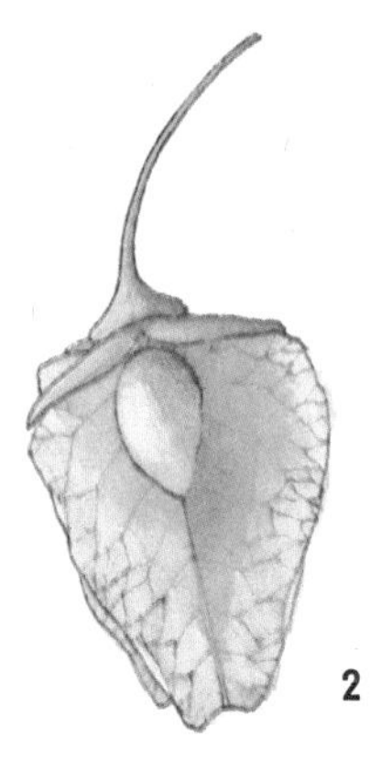

Great Water Dock is truly one of the largest of the lowland docks, with stems sometimes more than 2 m high and a large polycephalous rhizome. Everything about this dock is large, even the leaves — the basal leaves, which are flat and lanceolate, frequently measure 20 by 50 cm, but may sometimes be more than 1 m long! All the leaves have a prominent primary vein and tapering wedge-shaped base (1). The flowers are arranged in clustered pyramidal inflorescences and have a characteristic structure (2) described in greater detail in the meadow species *Rumex acetosa* L. or Common Sorrel (see page 64).

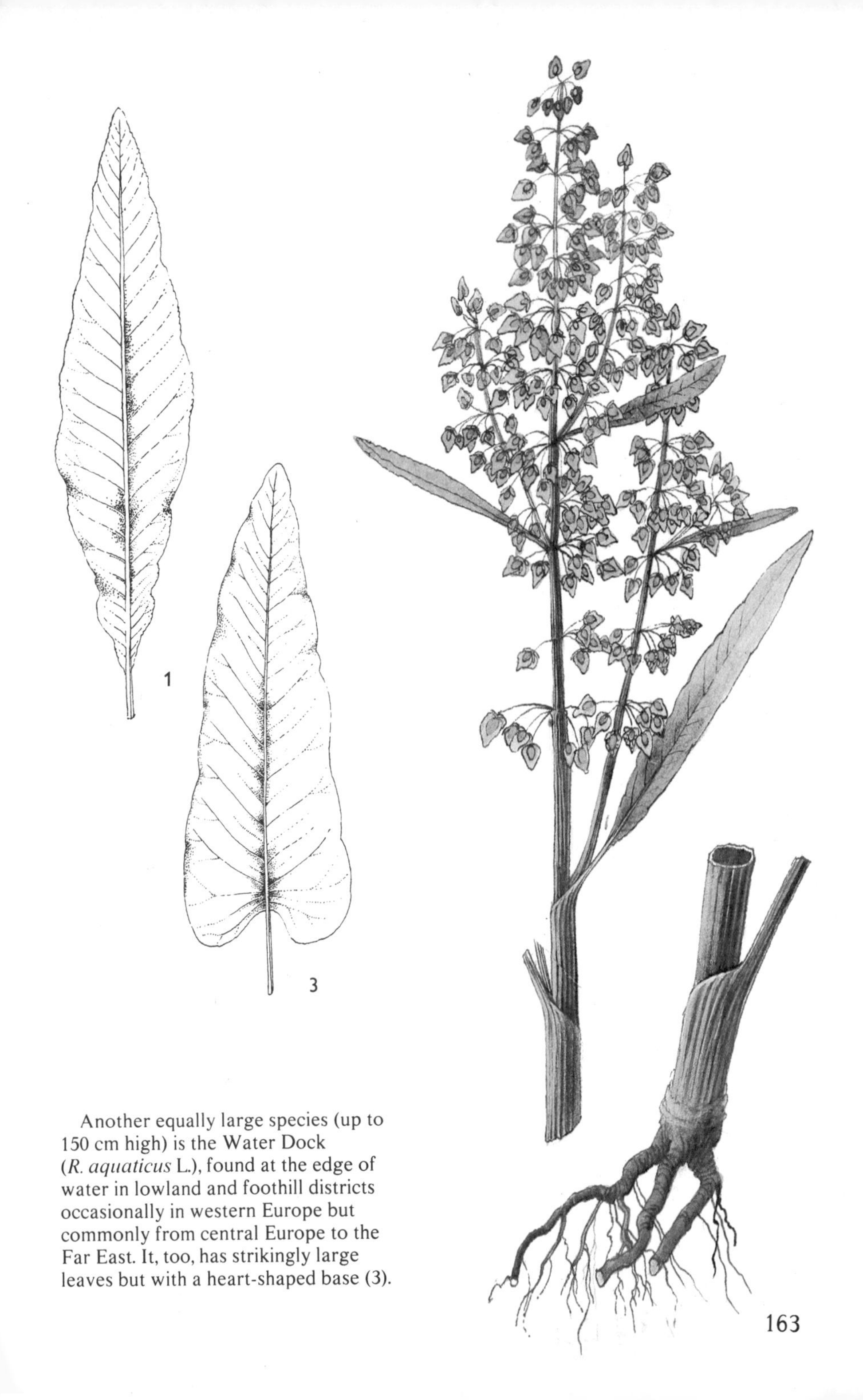

Another equally large species (up to 150 cm high) is the Water Dock (*R. aquaticus* L.), found at the edge of water in lowland and foothill districts occasionally in western Europe but commonly from central Europe to the Far East. It, too, has strikingly large leaves but with a heart-shaped base (3).

Common Spike-rush
Cyperaceae

Eleocharis palustris (L.) Roem. & Schult.

Spike-rushes are endearing marsh plants, which, when in a large mass, look like regiments of midget warriors in battle array with spears raised – a very pretty sight particularly in the cracked mud of the exposed bottoms of drained ponds.

Eleocharis was established as a separate genus in 1810, but some of its species, including the Common Spike-rush, were described before that by Linné and classed in the genus *Scirpus.* Later scientists divided the relatively variable species *E. palustris* into several minor species.

Most spike-rushes are circumpolar in the northern hemisphere and almost cosmopolitan. This applies also to the three illustrated species, even though, for instance, North American populations of the Few-flowered Spike-rush (*E. quinqueflora* (F. K. Hartm.) O. Schwarz) are usually considered to be separate taxons and in the narrow sense this spike-rush is indigenous only to north-western Europe.

The three selected spike-rushes serve as examples of the different habitats in which these plants grow – all, however, with a common denominator: the vicinity of water. The Common Spike-rush grows on pond and river banks, in marshes and in wet meadows; the Ovoid Spike-rush (*E. ovata* (Roth.) R.&Sch.) is the commonest inhabitant and pioneer plant of exposed bottoms; and the Few-flowered Spike-rush is a rare plant of sandy shores and marsh meadows.

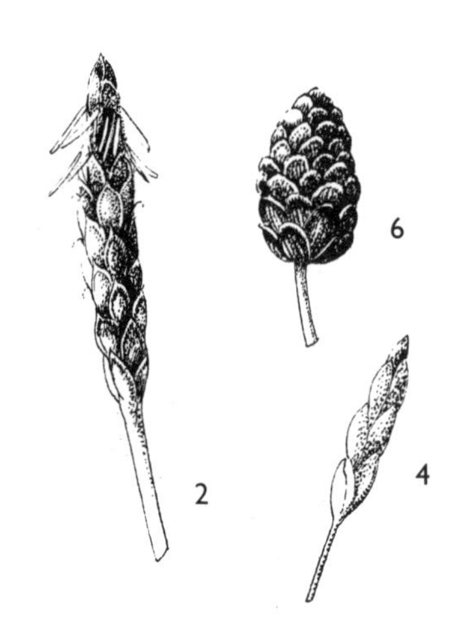

Spike-rushes are small, dainty plants, usually only 10–30 cm high and very much alike at first glance.

Common Spike-rush has a far-creeping rhizome and forms loose masses (1). It is a perennial with erect stems terminated by a spikelet composed of many hermaphroditic flowers (2).

Few-flowered Spike-rush (*E. quinqueflora*) is also a perennial herb but slightly tufted and slender (3) with small stems terminated by scanty spikelets composed of only a few flowers (4).

Ovoid Spike-rush (*E. ovata* (Roth.) R.&Sch) is a densely-tufted annual herb (5) with fibrous roots and blunt, broadly-ovoid spikelets (6).

1
5
3

Cowbane, Water Hemlock Umbelliferae
Cicuta virosa L.

Cowbane is one of the most poisonous waterside plants. It grows scattered at the edge of still and slow-flowing water and on the exposed bottoms of drained ponds. It is more or less tied to acidic soils deficient in lime and as a rule also attains the largest proportions in acidic iron-rich soils.

Cowbane is native to the waters and marshes of central and northern Europe and northern and central Asia, its distribution extending to Kashmir and Japan. North American populations are either classed in a separate species (*C. occidentalis* Greene) or as a subspecies of *Cicuta virosa.*

The plant, particularly the tuberous base of the stems, contains the alkaloid cicutoxin of the group of convulsive poisons. It is absorbed very rapidly, the first symptoms of poisoning occurring within a few minutes after ingestion. These consist of acute pains in the abdominal region and retching followed by convulsions. This is accompanied by foaming at the mouth, grinding of teeth, ashen colouring of the skin and paralysis of the nerve centres. Treatment is difficult because there are no known antitoxins. Similar cases of poisoning were observed in animals (cattle, horses) and even in fish when the top parts of the plant were leached in the water.

The best means of identifying this dangerous plant is by the tubers, which in cross-section have transverse cavities (1). It is practically impossible to mistake this for some other plant if the stem tubers are sufficiently thickened. A word of warning, however! The knife used to cut the tuber must be thoroughly washed right away because even that might cause mild poisoning if it is used, for example, to spread butter on bread.

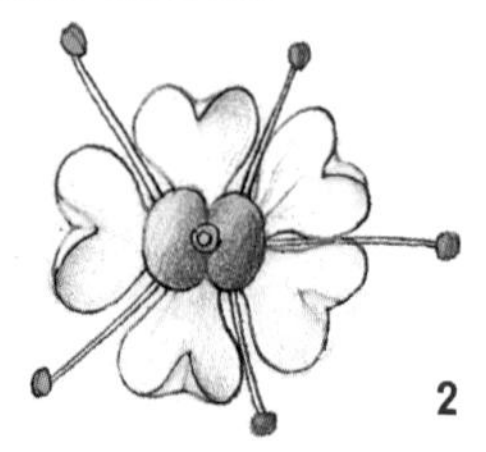

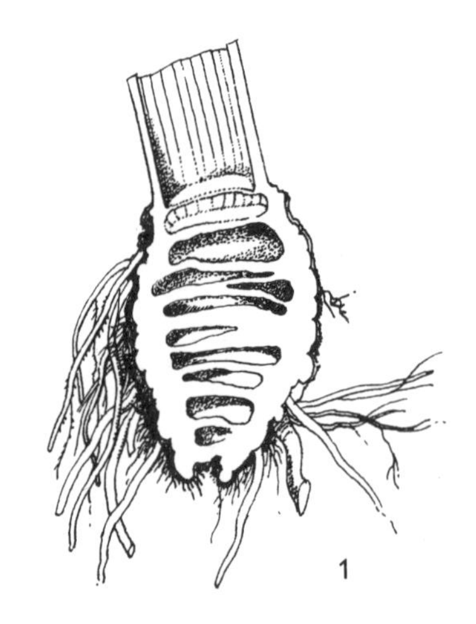

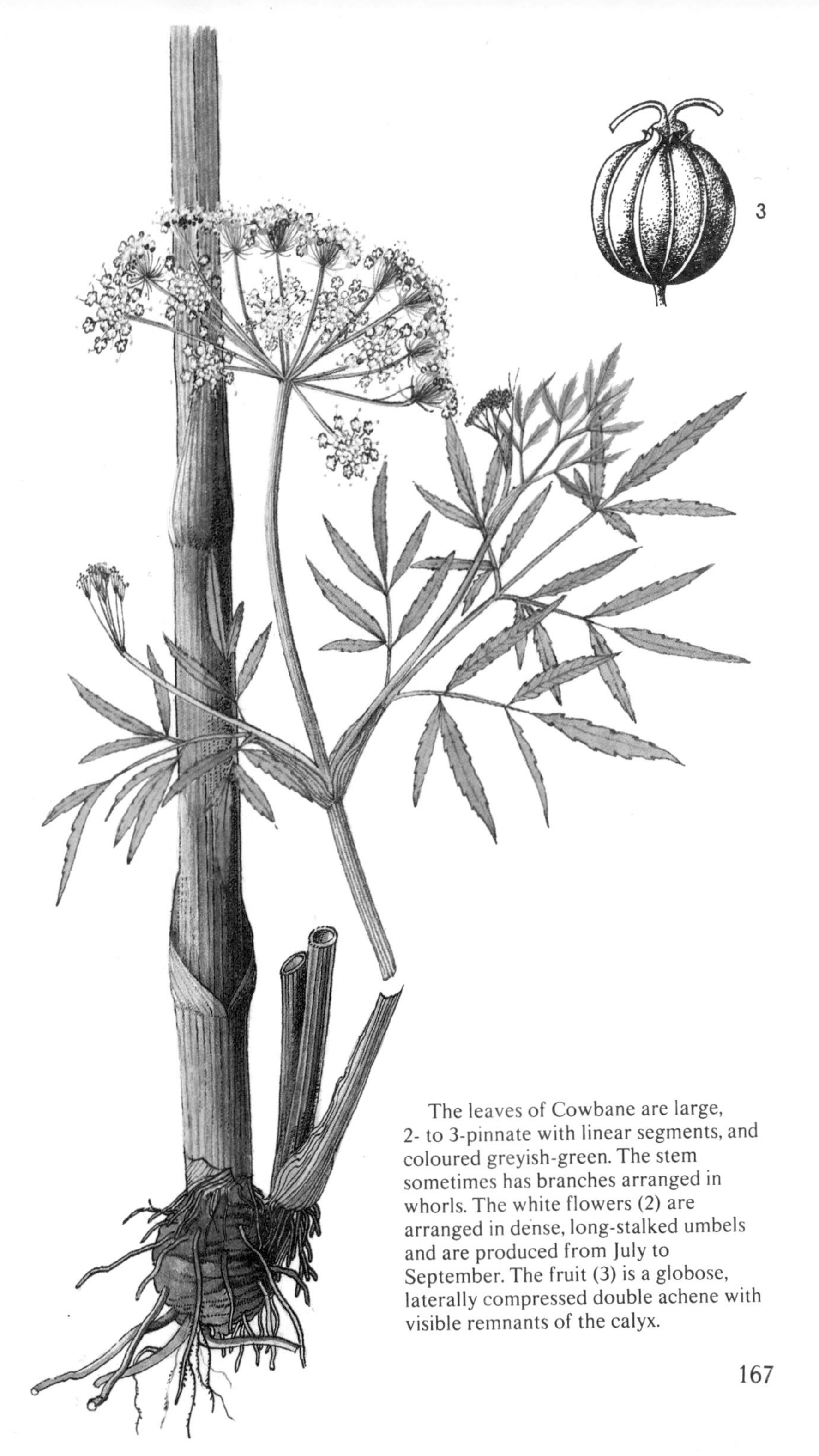

The leaves of Cowbane are large, 2- to 3-pinnate with linear segments, and coloured greyish-green. The stem sometimes has branches arranged in whorls. The white flowers (2) are arranged in dense, long-stalked umbels and are produced from July to September. The fruit (3) is a globose, laterally compressed double achene with visible remnants of the calyx.

Common Reed
Phragmites australis (Cav.) Trin.&Steud

Gramineae

Reed is probably the most important grass in the world next to sugar cane, cereal grasses and maize. It was used already by ancient civilizations and serves man to this day — as a source of cellulose and building material processed in modern factories near large river deltas (the Danube delta is particularly noted for this).

Reed is an example of a thoroughly cosmopolitan plant, which doubtless goes hand in hand with its great adaptability to various environments. Sometimes this robust marsh grass may be found growing in a dry rock crevice, in which case it is barely 30 cm high. The reed is thus also an example of great ecological variability — sometimes even the appearance of a single specimen changes during a single growing season. The ability to thrive even outside of water is made possible by the reed's large root system. Its rhizomes often penetrate the soil to the underground water level. These rhizomes may be up to 5 m long and always succeed in finding a suitable place to root. Old literature stated that in Lake Constance, Germany, reeds encroached upon the surface of the lake by as much as 3 m in a single year. Also known are so-called 'floating islands' — dense masses of tangled reed rhizomes and decaying stems plus humus in which even other, taller plants may be growing.

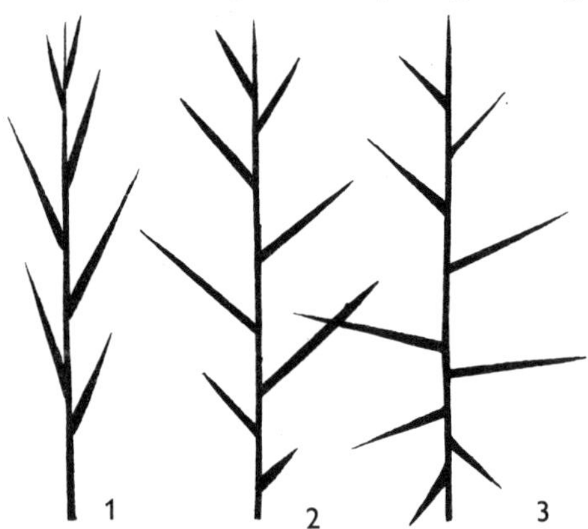

Common Reed is a robust, 1—5-m-high grass (though it may even reach a height of 10 m!) coloured greyish-green with large, long rhizomes and numerous shoots. The leaves on the stem are in two rows and their position is characteristic for the various seasons of the year: at the beginning of the growing period they form an acute angle with the stem (1), in July and August they gradually spread out until they are almost horizontal (2, 3). Because reed-beds are often exposed to winds they acquire a characteristic appearance with leaves and spent panicles all pointing to one side in the direction of the prevailing winds (so-called 'wind-trained' or 'flag-form' beds).

The tissues of reed are rich in silicic acid which, though it decreases the value of older plants as fodder, makes them useful as building material.

The flowers are arranged in large panicles composed of 3- to 8-flowered spikelets; the bottom floret is always male, the others hermaphroditic. The flowering period is from July to September.

Gipsy-wort
Lycopus europaeus L.

Labiatae

Of the 11 species of the relatively small genus *Lycopus* five are Eurasian, five are native to North America, and one is indigenous to Australia. The North American *L. virginianus* L., known as Bugle-weed, greatly resembles the European Gipsy-wort and, like the latter, was formerly also a medicinal plant called *Herba Lycopi virginiani.*

The top parts of Gipsy-wort cut during the flowering period yield a valuable drug with cardiotonic effect (one that affects the action of the heart). It calms heart palpitations and influences sleep as well as the function of the endocrine glands. Though it is innocuous even if used for a longer period it should be taken with the approval and under the supervision of a physician. The top parts were formerly also used to treat Basedow's disease (bronchocele) and in medieval days to treat fever and haemorrhage. The glycoside lycopin, contained in the drug, was used to treat malaria until recently. Gipsy-wort is a useful plant in other ways as well. The fresh juice pressed from the plant and mixed with copperas makes an excellent black dye.

Gipsy-wort grows in waterside thickets and damp alder groves and is a regular component of reed beds and shoreline vegetation on the margin of still and flowing water, but is not present in large numbers.

Gipsy-worts are perennial herbs with erect, hollow, four-angled stems sometimes more than 1 m high and usually covered with short down.

Lycopus europaeus is a very variable herb, with marked differences between plants growing in drier conditions and those growing in damp (flooded) situations. The leaves on a single plant are also of different kinds: those at the bottom are more deeply lobed with nearly linear segments (1), those at the top are more shallowly lobed with deeper clefts only at the base (2). The flowers of

Gipsy-wort (like those of Mint) are nearly regular, four-lobed, though it is a labiate (i.e. lipped) plant (3). They are stalkless and grow in clusters of 10 to 20 from the axils of the opposite leaves.

The flowering period is in late summer, from July to September.

Tripartite Bur-marigold
Bidens tripartita L.

Compositae

Tripartite Bur-marigold is a Eurasian species distributed throughout practically the entire continent. It tolerates the rugged climate at 65° latitude North and is also found (though only very occasionally) in the Mediterranean. One of its varieties also grows in south-western Asia. Occurrences in Africa are rare and in Australia and North America it is not an indigenous plant but was introduced there. It is found chiefly in lowland ponds and lakes but ascends alongside water even into mountains (in the Alps to the elevations of approximately 1,700 m). It is a very adaptable species so that it spreads even via such unusual means as railway drainage ditches.

A common species in North America is *Bidens frondosa* L. which was brought to Europe in the late 18th century (in 1792 it was cultivated as an exotic plant at Montpellier in France). During the ensuing two centuries, however, it spread so extensively that nowadays it is very common in Europe. In Czechoslovakia, for instance, it spread in the course of 20 years (1948—1968) from the Labe River system far into the upper reaches of the other Bohemian rivers.

Bur-marigold spreads readily thanks to its profusion of fruits furnished with setaceous bristles ('teeth') that catch in the fur of animals, plumage of birds, and man's clothing. These fruits also float on water, remaining on the surface for as long as one month, with some 80 per cent of the achenes (cypselas) falling to the bottom over a period of eight months, a period long enough for them to be carried a goodly distance by the current.

Bur-marigolds are annual, sometimes tufted herbs, marked by great fertility. Their name is derived from the Latin words *bis*, meaning two, and *dens*, meaning tooth, and refers to the prominent setaceous bristles ('teeth') — in reality a modified pappus — on the

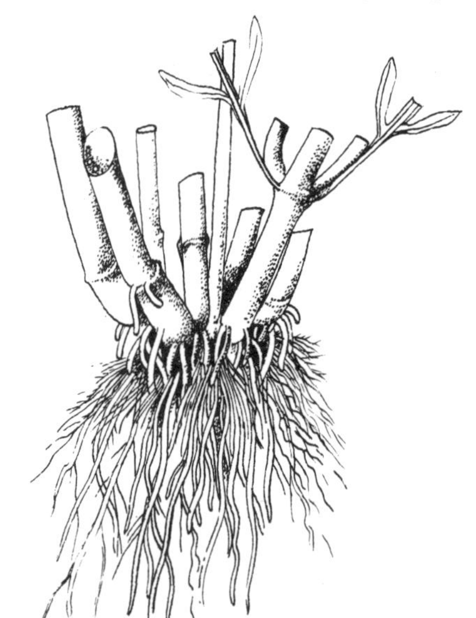

fruits (cypselas). The individual species can even be distinguished by the characteristically shaped fruits: the cypselas of Tripartite Bur-marigold are plumper, with two to three 'teeth' (1), those of the North American species *B. frondosa* are more elongate and do not have deflexed setae on the edges of the fruit (2). Most of the leaves of Tripartite Bur-marigold are odd-pinnate (2- to 5-partite) and opposite (3), the uppermost leaves are sometimes undivided (4).

The yellow ray flowers in the globose heads (5) are usually sterile, the central disc flowers are hermaphroditic. The flowering period is in late summer, from August to October.

Common Water-starwort
Callitriche stagnalis Scop.

If you see masses of 'stars' on the surface of a still backwater or stream, it will mean you have found a starwort of the genus *Callitriche.* The systematic classification of water-starworts and determination of their relationship to other plants is very difficult. The order Callitrichales is composed of a single family, Callitrichaceae, consisting of a single genus *Callitriche;* a further singularity is that the flowers are unisexual and the male flowers are reduced to a single stamen.

Water starworts are tiny, dainty and extremely variable plants forming aquatic, submerged or floating forms as well as terrestrial forms growing for a brief period in wet mud.

Common Water-starwort is most common in shallow slow-flowing or still water and less common in wet soil. In Europe it has been found primarily in the western and central regions. Far more common is the Spring Water-starwort (*C. palustris* L.), considered to be an arctic-alpine element found, besides central Europe, in Scandinavia and Iceland. In England it occurs only in a single locality. It stands up well to cold and becomes green even under ice.

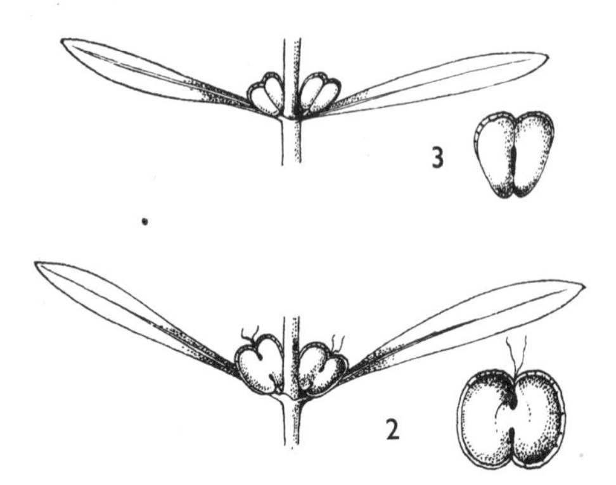

Water-starworts are dainty, fresh-green herbs, which may be perennial (those that grow in water) or annual (terrestrial forms), with narrowly-linear leaves clustered at the top of the stem in star-like rosettes that are particularly striking viewed from above (1).

The minute unisexual flowers are greatly reduced; the male flowers, as has already been stated, have only a single stamen, the female flowers only a single pistil with two styles and four ovaries that develop into four-seeded fruits after pollination. The perianth segments are reduced to two scales. The fruits of

C. stagnalis are broadly winged with short styles, if these remain on the plant (2). The fruits of *C. palustris* have only slightly developed wings at the apex and are without styles, for these soon fall (3).
The flowers are borne singly, from April to October.

1

Water Violet Primulaceae
Hottonia palustris L.

Water Violet is the only member of the primrose family with leaves divided into thread-like segments (comb-like). This evolutionary deviation was caused by the aquatic environment in which it grows. Such adaptations of the leaf blades have been seen in a number of other submerged plants illustrated in this book.

Hottonia palustris grows scattered in pools and shallow margins of ponds as well as in oxbow lakes. It often occurs even in the undergrowth of periodically flooded alder groves. When the water level drops it may even occur as a terrestrial form. Such plants have a greatly shortened stem and pinnate leaves clustered as if in a ground rosette. Viewed from above they are very decorative.

Water Violet often spreads by vegetative means, with 'buds' that overwinter in the mud. At other times the whole plant rips off from the muddy bottom before flowering and rises to the surface, where it remains until late summer before rooting again in the mud.

Water Violet is distributed in the temperate regions of Europe from England and southern Sweden to the Urals. Aquarists in Europe sometimes grow the North American species *H. inflata* Ell. (with a markedly flattened stem), native to the eastern United States.

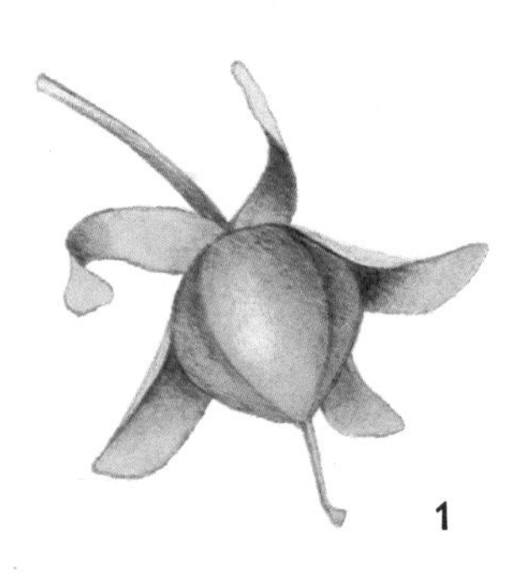

Water Violets were given their scientific name in honour of P. Hotton, Professor of Botany at Leiden. They are perennial herbs, 20—60 cm high — some growing in deep water were found to be more than 1 m long — with branched stems. The leaves are alternate, sometimes arranged in seeming whorls, and finely divided into thread-like, pointed segments with white thread-like rootlets occasionally growing from the axils.

The flowers are in whorl-like racemes on 10—30-cm-high scapes. As in other plants of the primrose family they are distinguished by dimorphism (heterostyly). After pollination the scape bearing the maturing fruits (1) submerges. The flowering period is from May to June.

Canadian Waterweed Hydrocharitaceae
Elodea canadensis Michx.

Elodea canadensis is called 'water plague' in some languages and no wonder. The first plants were brought from their native North America to Europe in the late 1830s. Round about the middle of the century, when they were still popular, they were an article of exchange between Europe's botanical gardens. About a hundred years ago they appeared in central Europe but within far less time than that they had infested countless ponds and bodies of water. Though those that had crossed the sea to Europe were only female plants incapable of producing fruits and seeds, waterweed's spread was rapid. All that was needed was for the floating stems to break up into small pieces which soon developed into new, sometimes several-metre-long shoots forming thick, tangled, impenetrable masses — a formidable obstacle not only to fishermen but also to navigation.

Waterweed thrived in shallow as well as deep water. Fresh plants are good green fodder comparable to clover. It is also grown in aquariums together with the related Dense-leaved Pondweed (*Egeria densa* Planch.), native to the subtropical regions of South America, whence it was introduced via North America to other continents. It has been encountered even in certain river canals in Germany.

Elodea canadensis is a submerged perennial aquatic with a jointed, much-branched, densely-leaved stem. The leaves, mostly three in a whorl (1), are sessile and imperceptibly serrate on the margin. The female individual introduced to Europe (Europe's entire present population is probably a clone!) flowers very briefly. The flowers are minute and practically invisible from the shore.

The related, introduced, Dense-leaved Pondweed (*Elodea densa*) has prominent, gleaming white female flowers (2) and leaves mostly five in a whorl.

Flowers are produced only rarely, from May to August.

1

Mare's-tail
Hippuris vulgaris L.

Mare's-tail (*Hippuris*) is the only genus in the family Hippuridaceae of the isolated order Hippuridales; its relationship to other groups of angiosperms is very hazy. However, it definitely has nothing in common with horsetails, even though there is a slight resemblance. It seems that this is a very old branch from the evolutionary aspect, one which existed perhaps as far back as the Mesozoic era, in other words about 140 million years ago. Fossil remnants of these plants were found also in sedimentary deposits from interglacial periods — the Quaternary era. The several existing species are found in the northern hemisphere and South America — one being the illustrated *H. vulgaris.*

They grow in still or slow-flowing water in mud and may form continuous masses that look quite exotic. Plants that catch a foothold in more rapidly flowing water are flag-form and permanently submerged.

Mare's-tail stands up well to low temperatures — it grows even in parts of South America influenced by the Antarctic and in arctic regions in the north. The illustrated species is a typical inhabitant of northern (arctic) lakes where it grows on the sparsely vegetated shores together with a few other species such as Marsh Cinquefoil (*Potentilla palustris*).

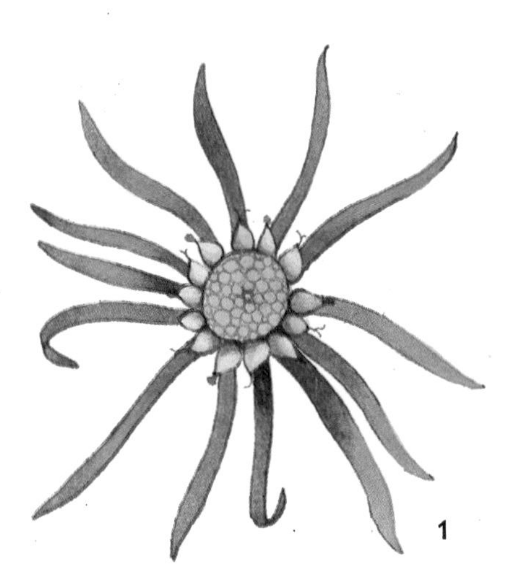

Mare's-tail is a perennial aquatic and marsh plant with long, creeping, rooting rhizome and erect stems about 40 cm high. Floating stems or ones growing in rapid-flowing water may be up to 1 m long. The stem is 'jointed' — an impression created by the thick whorls of hair-like leaves. Reduced flowers form in the axils of the upper leaves that are above water. They are unisexual, either male or female, and are composed of only a single stamen or single pistil. The floral envelopes, i.e. the calyx and corolla, are atrophied, nonexistent. The position of the flowers and arrangement of the leaf whorl is best seen on the

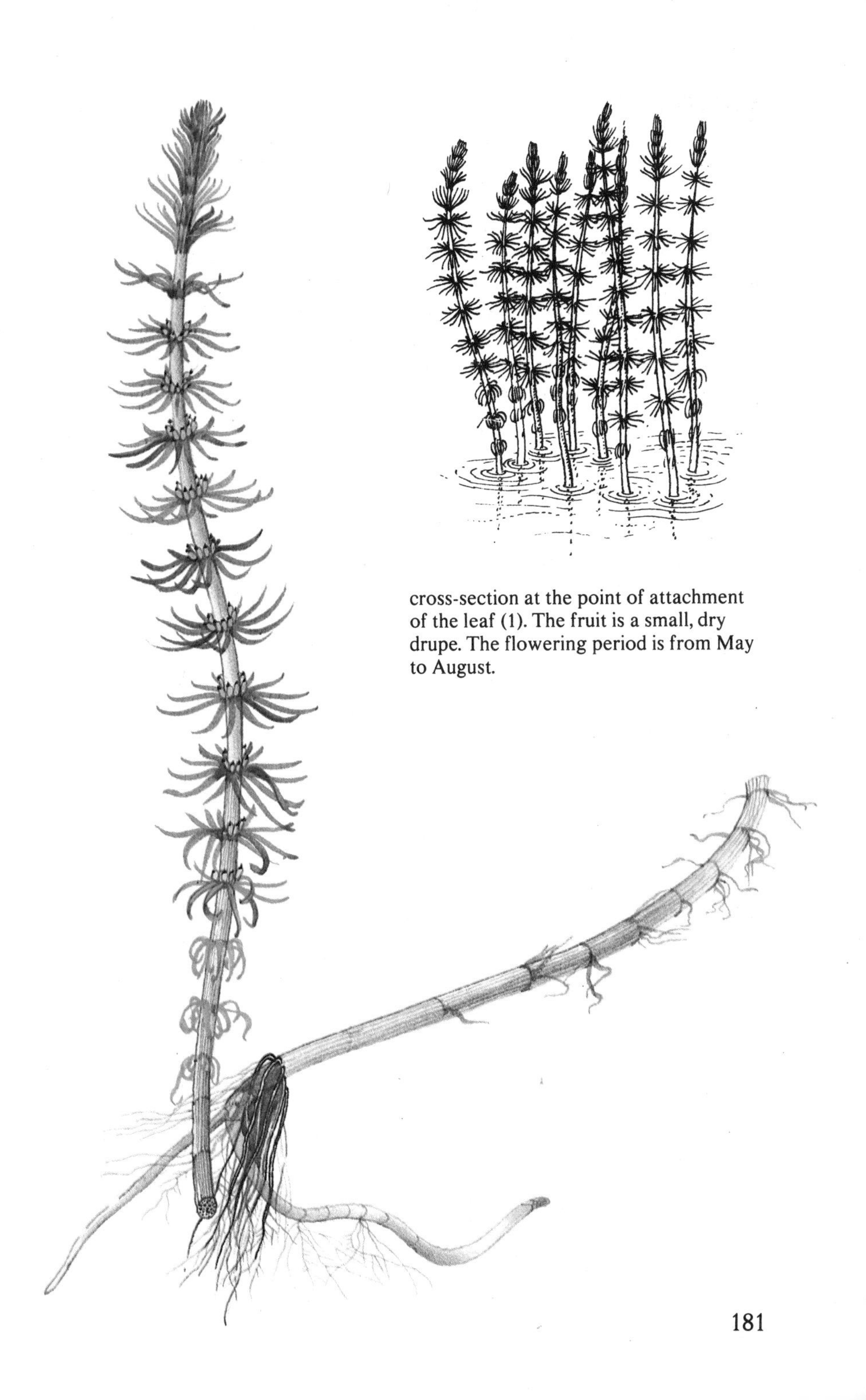

cross-section at the point of attachment of the leaf (1). The fruit is a small, dry drupe. The flowering period is from May to August.

Common Quill-wort *Isoetes lacustris* L.

Quill-worts are probably the oddest of the plants presented in this book and among the most peculiar of all plants on the Earth. They are an isolated group of Pteridophytes found solely on lake beds or the bottoms of other bodies of water. Together with the South American genus *Stylites* they are the last living remains of the almost extinct evolutionary branch of lycopods (Lycopodiophyta). This ancient group had its golden age in the Carboniferous period of the Paleozoic and its remnants must be sought in black coal seams. The large terrestrial tree-like genera have been extinct for ages and existing descendants number some 70 species, found chiefly in North America and the Mediterranean region. North-western Europe has only two surviving species occurring in isolated localities also in central Europe. They were still relatively common in Europe at the end of the last glaciation but nowadays their refuges are glacial lakes where they can grow even at great depths — to 5 m and more. That is why they are termed relicts.

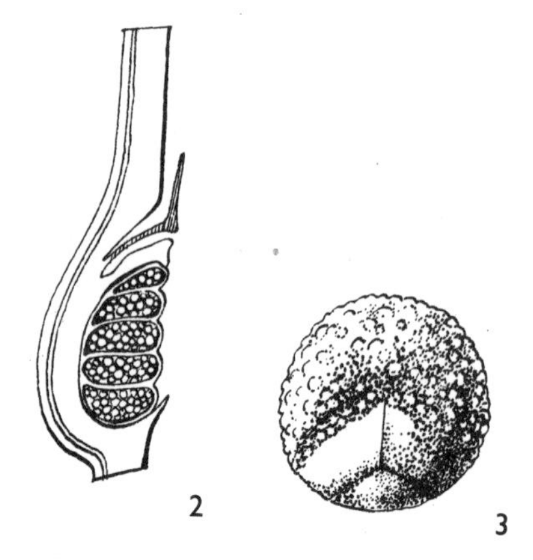

The scientific name of Quill-worts is derived from the Greek words *isos*, meaning the same, and *étos*, meaning year, because they look the same throughout the year.

Quill-worts are tufted herbs with narrowly-linear, quill-like leaves (1) which have sporangia embedded in the widened leaf-base and above these a membranous appendage called a ligule (2). The spores consist of megaspores, which develop into female organs, and microspores, which develop into male spermatozoids.

The two species *I. lacustris* and *I. echinospora* Durieu differ only slightly in the minute surface structure of the megaspores (3) and in their size, leaf shape (fig. 4 shows a cross-section of the

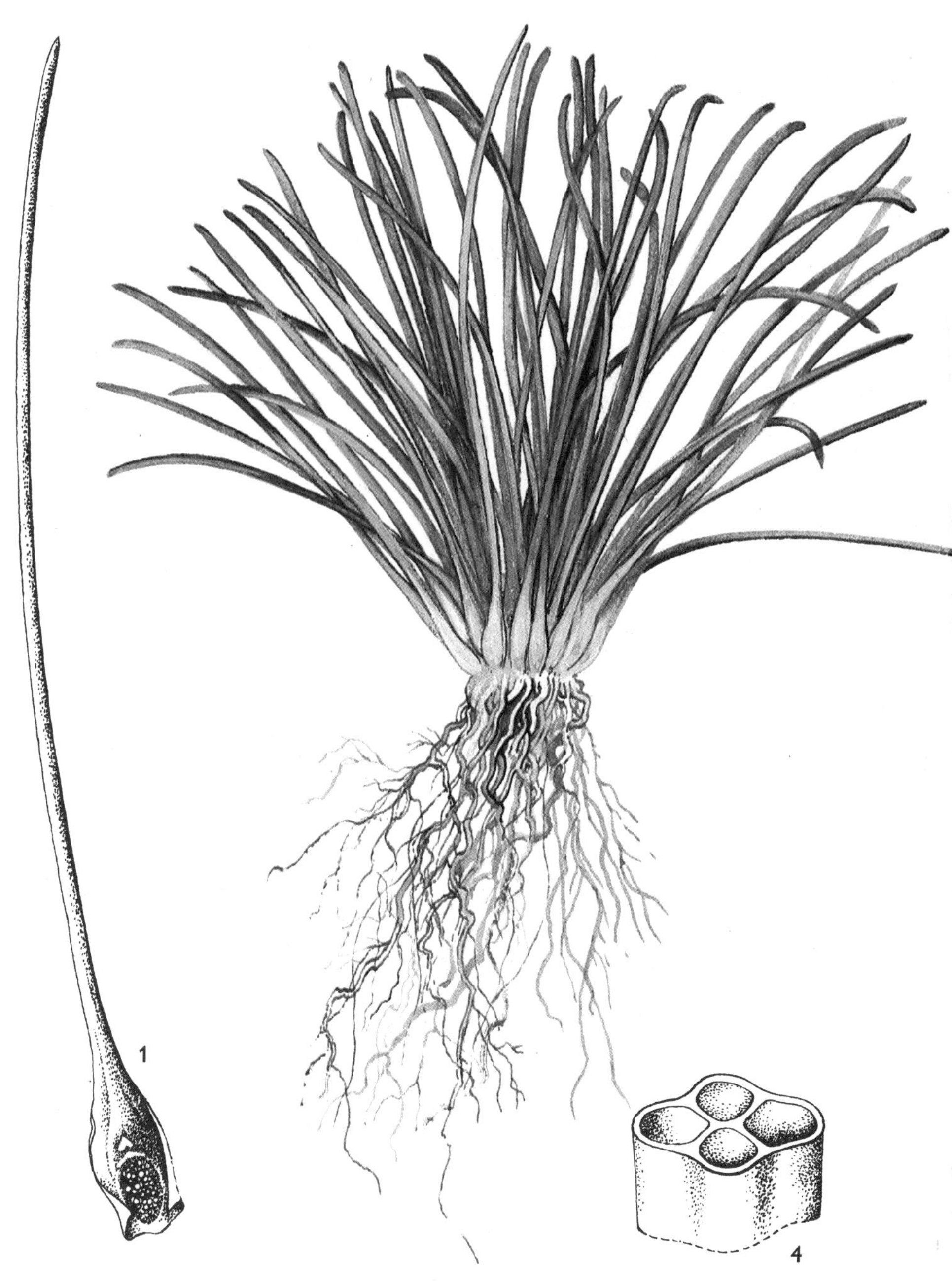

leaf of *I. lacustris*) and in coloration (*I. echinospora* is a lighter green). It is difficult to determine the time of 'flowering'. The spores are released from the sporangia from approximately May to September.

Broad-leaved Pondweed *Potamogeton natans* L.

Potamogetonaceae

Selecting a suitable representative of the large genus *Potamogeton* (which numbers 90 species) is a difficult matter for each, in its way, is interesting. The pondweeds are a widespread family distributed from the tropical to the temperate zone. They are very adaptable to the environment in which they grow and very variable, particularly in the shape of the leaves. The genus *Potamogeton* includes among its members species with broad, floating leaves (Broad-leaved Pondweed), species with thread-like leaves reminiscent of fennel (Fennel-leaved Pondweed — *P. pectinatus* L.) and species with long linear leaves (Shetland Pondweed — *P. rutilus* Wolfg.).

They are generally submerged or floating aquatics firmly rooted in the muddy bottom. The leaves and stems contain ample aerating tissues which supply the sumberged parts with air. They also multiply readily by vegetative means for parts that are broken off root rapidly. Water birds also help spread the plants: the seeds of pondweeds are indigestible and thus are deposited with the bird's faeces sometimes in distant and isolated bodies of water.

Masses of pondweed are welcomed in pisciculture for they provide nurseries with a plant where fish can lay their eggs. For many swimmers, however, submerged colonies of pondweed are most unpleasant.

Potamós and *geitón* are the Greek words for river and neighbour or close and 'potamogeiton' was the ancient term for certain aquatic plants which is nowadays only applied to pondweeds. They are perennial herbs with rhizomes firmly anchored in the mud and long, floating stems. The leaves of Broad-leaved Pondweed (1) are up to 12 cm long and 6 cm wide, long-stalked, leathery and float on the water. *P. fluitans* Roth. is an

4

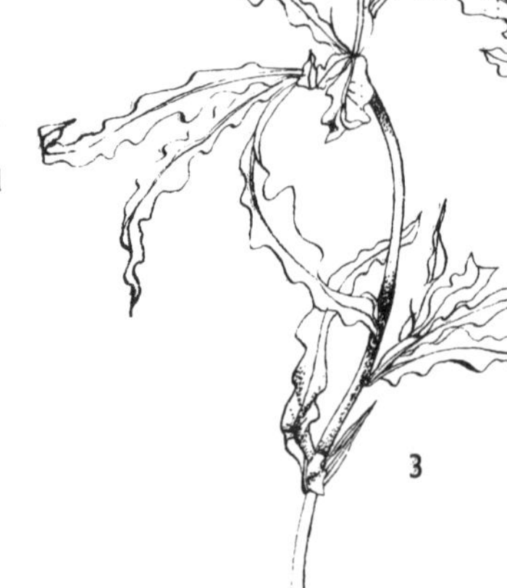

3

example of pondweed with two types of leaves: submerged leaves that are linear-lanceolate and floating leaves that are broader, ovate (2). The leaves of Curled Pondweed (*P. crispus* L.) are strongly undulate on the margin (3).

The flowers are borne from June to August in erect spikes above the water. They are hermaphroditic; the green connective fulfills the function of the perianth segments which are absent (4).

River Crowfoot
Batrachium fluitans (Lam.) Wimm.

Ranunculaceae

Crowfoots, when and if they flower, form a kind of white haze floating above the surface of water, an effect which is usually visually stunning. They form large masses in ponds and the calm spots in rivers where they are distributed locally.

River Crowfoot is found in more rapidly-flowing rivers, chiefly in the middle reaches. In the calm spots where it is rooted it forms dense undulating masses that shelter pike and eels. Water Crowfoot (*B. aquatile* Dum.), an important plant with a practically cosmopolitan distribution, prefers still water and is found in pools and oxbow lakes at depths of about 50 cm. On exposed bottoms and in the shoreline zone it sometimes occurs only as a terrestrial form with undivided leaves similar to the floating leaves.

Crowfoots are often classed in the genus *Ranunculus;* that they are a separate genus is borne out by recent studies, e.g. of their anatomy, morphology and floral biology.

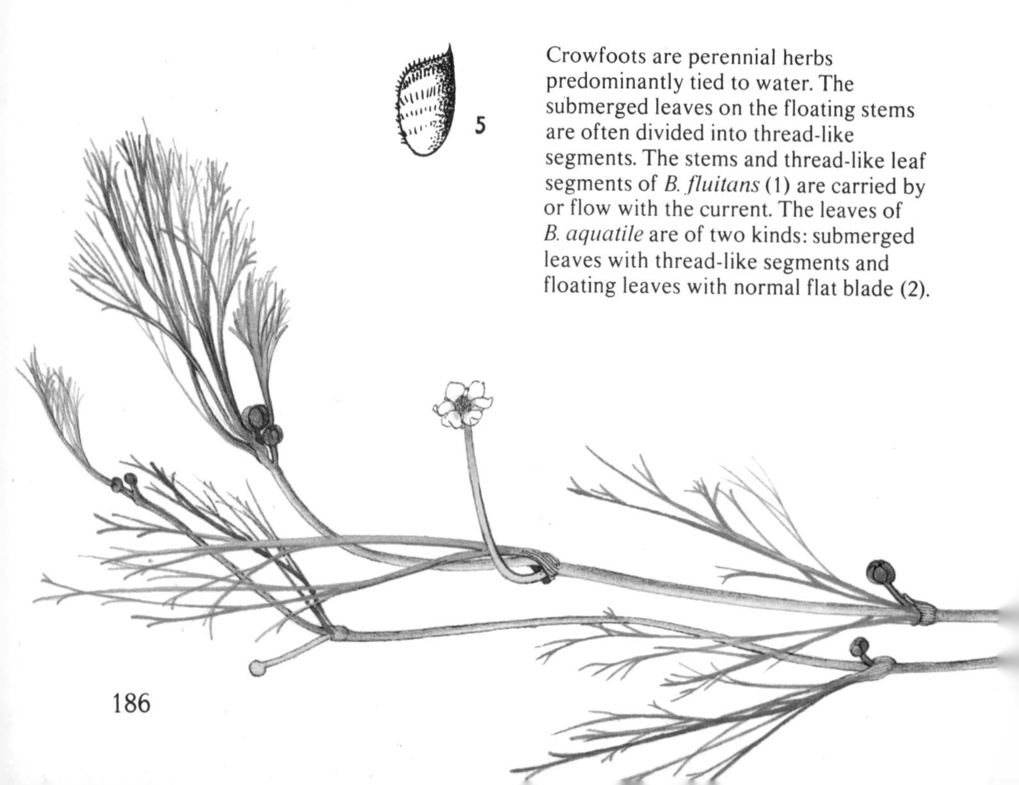

Crowfoots are perennial herbs predominantly tied to water. The submerged leaves on the floating stems are often divided into thread-like segments. The stems and thread-like leaf segments of *B. fluitans* (1) are carried by or flow with the current. The leaves of *B. aquatile* are of two kinds: submerged leaves with thread-like segments and floating leaves with normal flat blade (2).

The flowers are usually 5-merous (3), but they may sometimes have as many as seven petals. The flowers borne in June are conspicuously larger than those produced during the second flowering in August and September.

After pollination the flowers develop into a 'head' (4) composed of separate achenes that float on the surface (5).

Amphibious Bistort
Polygonum amphibium L.

Polygonaceae

Amphibious Bistort is a truly ubiquitous amphibian. It may pass its entire life-cycle in its aquatic form or in its terrestrial form. The two forms are quite different and may readily be considered to be two separate species by the layman. Like many plants that are closely tied to water it has a circumpolar distribution in practically the whole of the temperate zone, occurring in Europe, Asia and North America as well as in India, Mexico, etc. It is a common and useful plant in still and slow-flowing water. It does best in shallow bodies of water with sandy and muddy bottoms and in ponds of medium depth. It spreads from shoreline reed-beds even to deeper open water but does not grow in very deep water. Bistort is an important plant in that it provides fish with shelter and suitable places for spawning. Formerly, when the draining of ponds for the summer was more common, Amphibious Bistort which formed a second, terrestrial form on the exposed bottom survived the 'dry' period better than any other aquatic plant.

The name of the genus — *Polygonum* — is derived from the Greek words *polýs*, meaning many, and *góny*, meaning knee or joint in reference to the stems of these plants which have many conspicuous joints or 'knees'.

The aquatic forms of Amphibious Bistort (1) have long stems, often more than 1 m long, composed of segments (internodes) measuring up to 20 cm in length; they are often much branched. The upper stem leaves float on the surface and die in succession during the course of the year. The spike-like inflorescence is always completely above water; the flowers are coloured deep pink.

The terrestrial forms of Amphibious Bistort have erect stems up to half a metre high with short internodes (2—7 cm) and leaves growing from sheathing stipules covered with long hairs (2). They rarely produce flowers; the spike-like inflorescence is short, the flowers pinkish.

The flowering period is in June and July; there may be a repeated flowering later in the season.

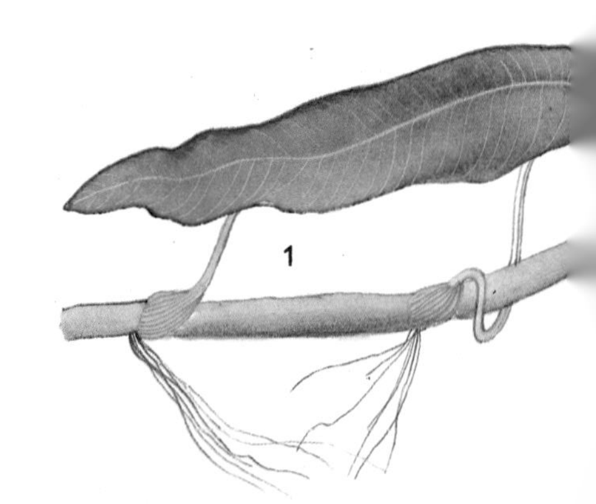

2

Frogbit
Hydrocharis morsus-ranae L.

Hydrocharitaceae

The freshwater Frogbit and Water Soldier (*Stratiotes aloides* L.) are something of an exception amongst the members of the frogbit family, which includes mostly marine plants. There are some nine genera and 31 species inhabiting the seas and, together with the members of the pondweed (Potamogetonaceae) family, they are the most common, and sometimes the only, gymnosperms in the marine flora.

Frogbit grows in still and slow-flowing water and in pools, small lakes and canals; it is most common in bodies of water with a small surface area where it may form thick, spreading masses. Because it produces numerous tillers such colonies become practically impenetrable, shading the water and making fishing difficult.

Frogbit serves as a good example of certain peculiarities of the root systems of aquatic plants. For example: because it is a floating aquatic the roots are freely suspended in the water and the root hairs remain on the roots throughout the plant's entire lifetime (unlike terrestrial plants that root in soil) thereby increasing the root surface about eight times!

Frogbit is a Euro-Siberian plant distributed chiefly in western and central Europe. It is absent in Scotland, northern Scandinavia and southern Europe.

The Latin name of the plant is something of a paradox, for the word *hydrocharis,* freely translated, means ornament of waters and *morsus-ranae* means frog pain or frog bite.

Frogbit is a perennial with short rhizome and long roots; the leaves float on the surface in characteristic rosettes. Non-flowering plants can be readily distinguished from the similar Fringed Water-lily *(Nymphoides peltata)* by the scaly stipules at the base of the leaf-stalks.

The 3-merous flowers of Frogbit are unisexual: the male flowers, with long stalks and 12 stamens, grow from the axils of membranous spathes; the female flowers are nearly sessile within the spathes and have six bifid stigmas (1). Frogbit often spreads by means of long stolons terminated by new leaf rosettes or ovoid, overwintering 'buds', which rise from the mud to the surface of the water in spring.

The flowers, if they are produced, are borne from June to August.

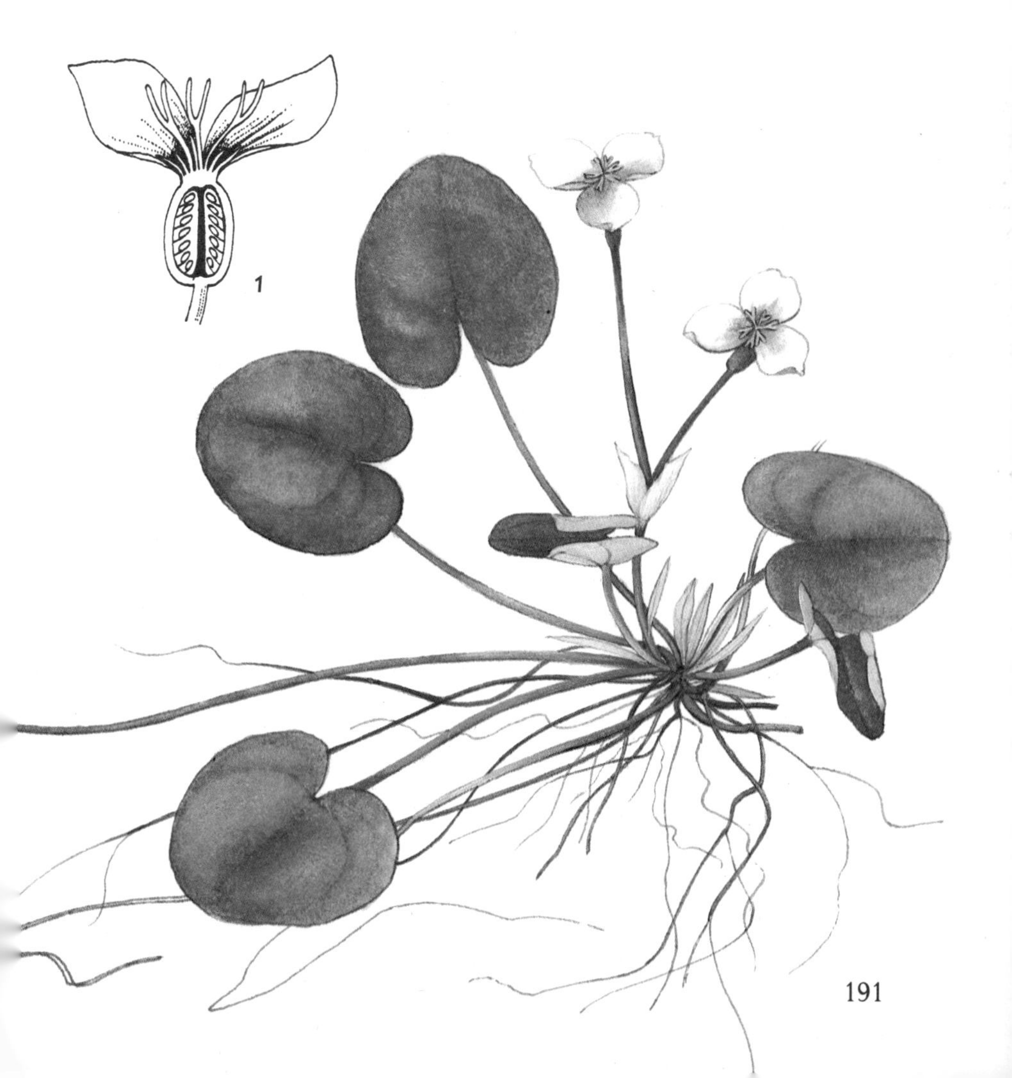

Yellow Water-lily, Brandy-bottle Nymphaeaceae
Nuphar lutea (L.) Sm.

Some fishermen and pisciculturists may not agree — but Yellow and White Water-lilies are among the loveliest of aquatic plants. Though they are still plentiful in places Yellow Water-lilies must alas be viewed as rare elements in nature. Their development is slow at first. The seeds, dispersed by water birds or by water, develop into seedlings which are fully grown and bear flowers for the first time only after about four years. The first leaves of the new growing season are formed already in the autumn of the preceding year. As soon as the temperature drops to 10°C, however, growth ceases and the plants prepare to overwinter. Growth is resumed in spring; thin submerged leaves develop first, followed by the characteristic, flat, leathery floating leaves.

Yellow Water-lilies grow best at depths of 80 — 200 cm in ponds, lakes, pools, backwaters and oxbow lakes. They are typical Euro-Siberian plants. Very rarely, however, do we encounter the Least Yellow Water-lily (*Nuphar pumila* (Timm.) DC.) nowadays. The construction of dams, pollution of rivers with industrial wastes, and the excessive fertilization of ponds have been responsible for the disappearance of this precious relict from Europe's flora.

The Yellow Water-lily was a popular herb of the herb healers of old, who used the drug obtained from the rhizomes as a sedative to treat nervous irritability. The drug contains certain alkaloids (nufarine, nufaridine) and a cardiotonic glycoside. Taken in larger doses it could cause paralysis of the respiratory centres.

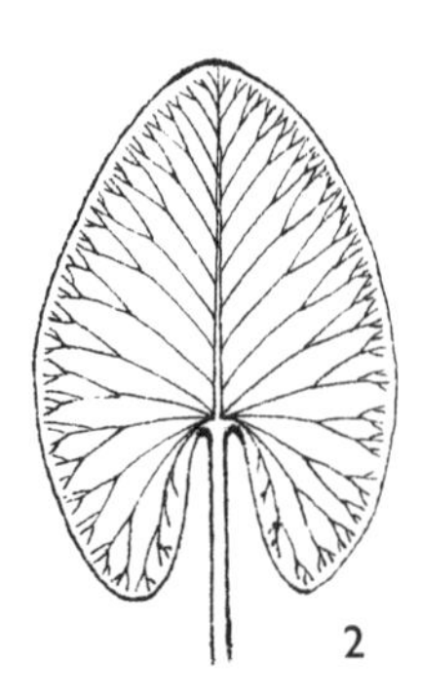

Yellow Water-lily is a perennial aquatic with a very thick and long rhizome (sometimes up to 10 cm thick and more than 2 m long), with roots growing from the underside and an upper surface that is irregular and marked by leaf scars (1). Growing from the tip of the rhizome are long-stalked leaves, some submerged and others, with typical forked venation (2), floating on the water. The blade may be up to 40 cm long and 30 cm wide.

The yellow flowers on long stalks are borne from June to August; they close at night. After pollination the pistil with funnel-shaped, lobed stigma develops into a long, multicapsular bottle-shaped fruit (3).

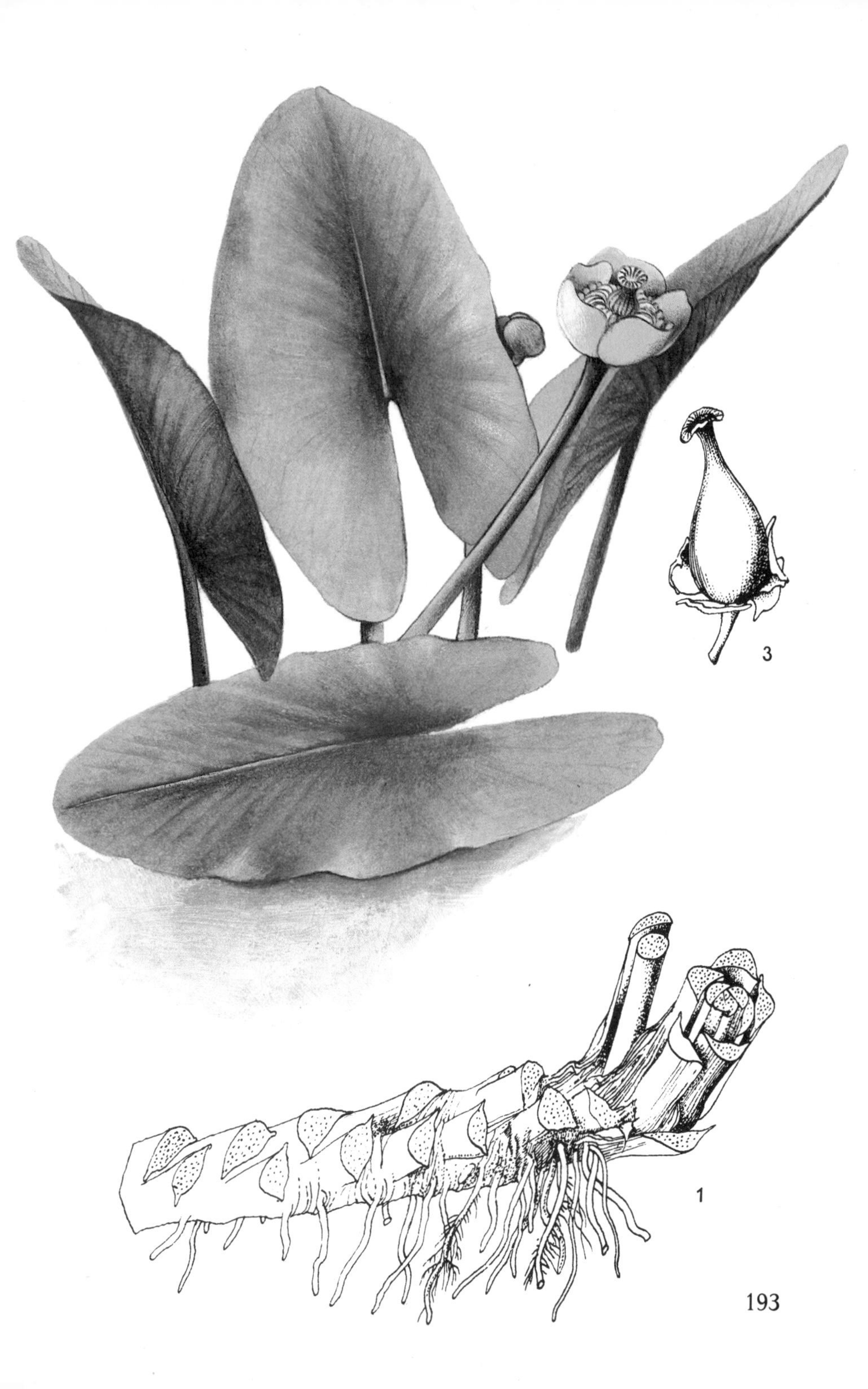
3
1

White Water-lily
Nymphaea alba L.

Nymphaeaceae

The White Water-lily is the subject of tales and legends, for instance, about how the morning and evening star fought over an arrow shot by an Indian chief until the sparks flew; these sparks fell into the waters on Earth and became water-lilies. Or else about enchanted nymphs and naiads that emerged from the blossoms on moonlit nights to dance on the water.

Today, in landscapes transformed by the inroads of civilization, White Water-lilies are already a rare species and in many countries are on the list of endangered plants. They grow in still waters with muddy bottoms in central and western Europe — *N. alba* is a European species and *N. candida* Presl a Euro-Siberian species with a more north-easterly distribution; the latter is absent in Great Britain and on the west-European coast.

The floating leaves of White Water-lilies are a model of this type of leaf, with pores located only on the upper side. The flowers open periodically, between 7am and 5 pm, on clear days. They are pollinated by insects, primarily flies. The petals fall after about a week and the flower stalk coils in a spiral and disappears below the surface. The fruit, a large, fleshy capsule like that of the poppy, matures underwater.

The rhizomes at one time yielded a drug used as a sedative to treat nervous irritability.

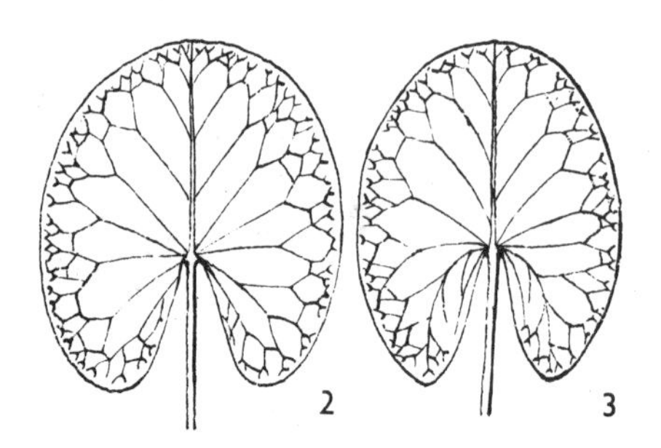

The Water-lily's name is also rooted in legend: one of the nymphs of Greek myth died of love for Hercules and was transformed into a flower, hence the name *nymphaea*.

White Water-lilies are perennial aquatics with a rounded rhizome about 5 cm thick and 30 — 50 cm long growing straight down in the mud (1) — unlike that of the Yellow Water-lily which grows at a slant. The leaf scars are distributed evenly round the periphery of

the rhizome and the roots are likewise distributed. The rhizome contains alkaloids and the glycoside nymphaein.

The leaves of White Water-lilies have a characteristic venation, differing from that of Yellow Water-lilies. The leaves of *Nymphaea alba* (2) have a deep and wide basal cleft, those of the Pure-white Lily (*N. candida*) (3) have a deep and narrow basal cleft, the lobes sometimes overlapping. The white flowers are up to 12 cm across, with numerous petals, and are borne on long stalks from June to August.

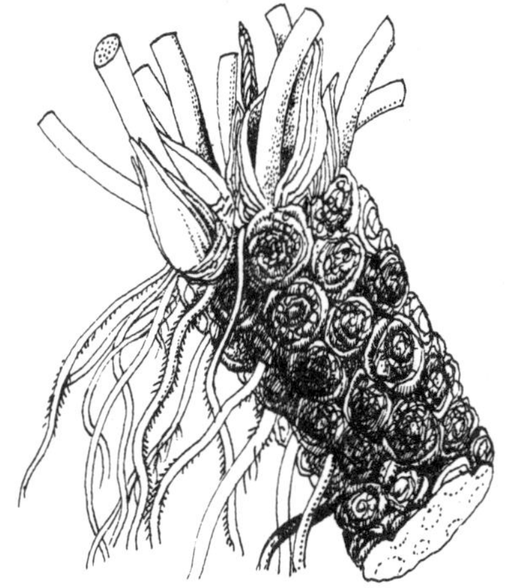

1

Common Bladderwort Lentibulariaceae
Utricularia vulgaris L.

Bladderworts, along with quillworts and sundews, are the most extraordinary of the plants depicted in this book. They are rootless aquatics (and thus cannot absorb food from the soil), with leaves adapted for trapping small water animals, that besides being able to supply their own food by normal photosynthesis can also obtain the nutrients they lack from the bodies of other organisms (mixotrophic). Located between the leaf segments are peculiar inflated bladders (modified leaves) with an opening closed by a kind of lid that only opens inward. Round the edge of the opening are spreading hairs ('antennae') that are sensitive to the touch. The instant a small water crustacean brushes against these antennae the lid opens abruptly and the whole bladder stretches, thereby sucking in water and with it the animal, which eventually dies in its prison. The substances released by the decomposition of its body are then absorbed by the plant's tissues.

Some naturalists assert that these bladders also serve as floats — before flowering they are usually filled with air so that the plants float near the surface and the flower stalks rise above the water.

Bladderworts are rare circumpolar plants of the northern hemisphere, where they grow scattered in still waters with muddy bottoms.

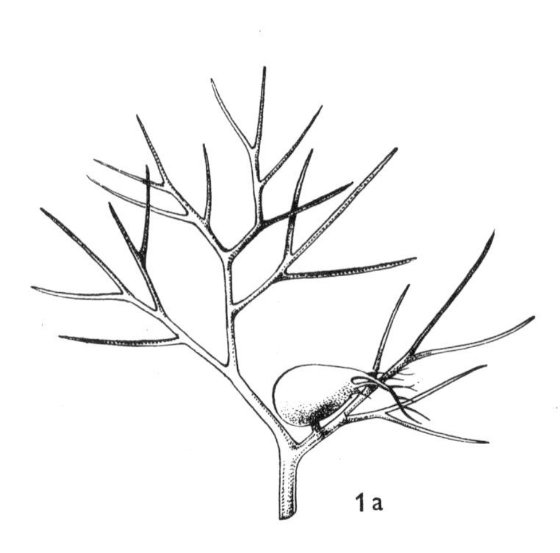

Bladderworts are delicate perennial aquatics without roots, though sometimes the non-green part of the stem becomes anchored in the mud — as in the Intermediate Bladderwort (*Utricularia intermedia* Hayne). The stems are sparsely branched with alternate leaves palmately divided into thread-like segments and bladder-like insect traps (1a, 1b) — the Latin word *utriculus* means small bag or pouch.

The flowers are arranged in a scanty raceme on a long stalk that rises above the water. The calyx is composed of two lobes, the corolla is bi-labiate, and the number of stamens, which are short, is likewise two.

Flowers are produced only rarely, from June to August. The plants generally multiply asexually, by means of buds that overwinter on the muddy bottom.

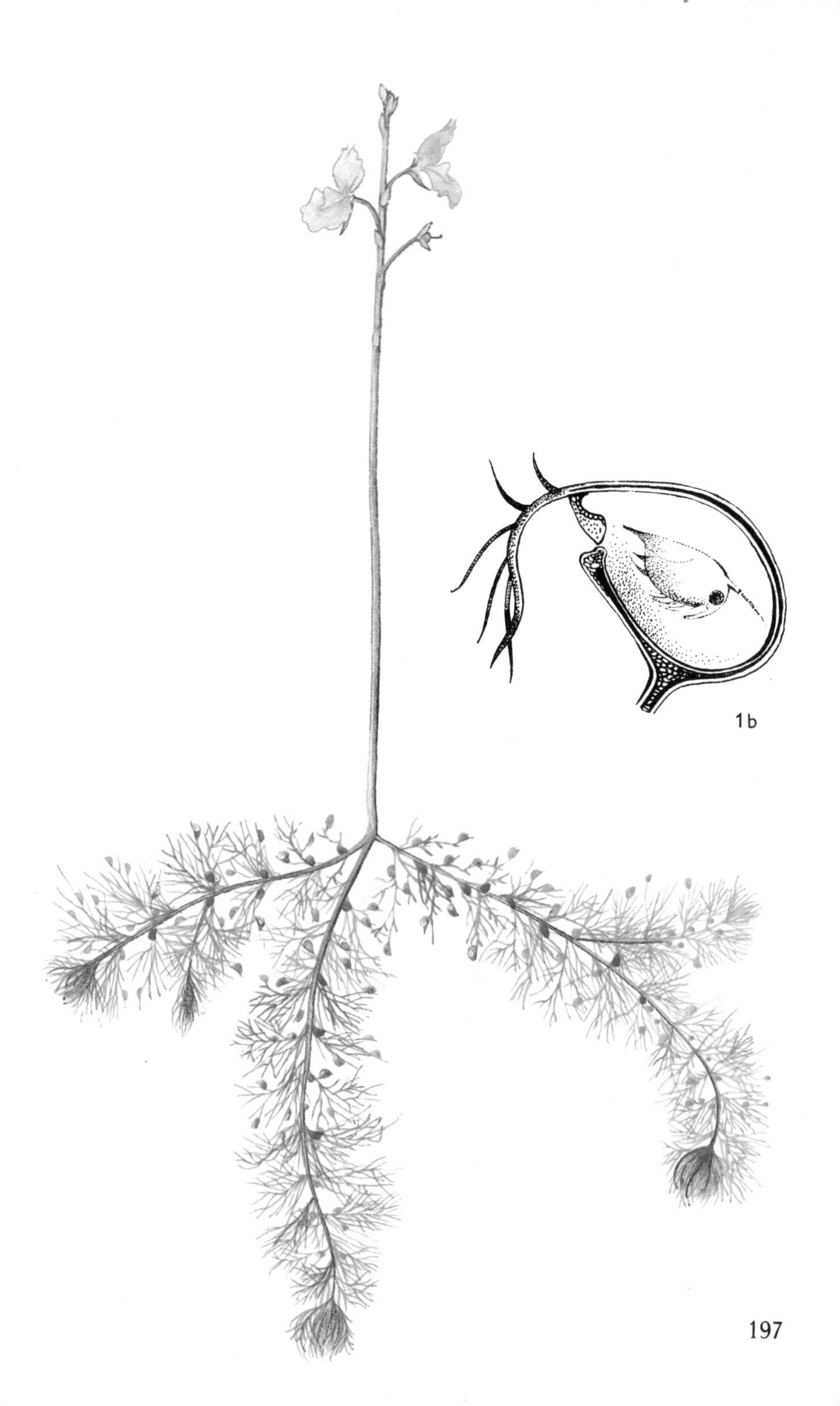
1b

Water Soldier
Hydrocharitaceae

Stratiotes aloides L.

The common name for this plant in some languages is Water Aloe, e.g. Wasseraloë in German and Faux aloës in French, and even the Latin name for the species, *aloides,* calls to mind the striking resemblance between the leaves as well as the whole leaf rosette and those of certain plants of the genus *Aloë.* Water Soldier, however, is definitely aquatic and its leaf rosettes open just below the water's surface. In deeper water one may sometimes also encounter submerged forms with more finely-leaved rosettes anchored in the bottom. Ordinarily, however, the half-submerged rosettes are free-floating, their long fleshy roots only occasionally anchoring in the mud. The plants are very adaptable and can be grown as ornamentals in stone troughs filled only with water without a bit of soil or sand on the bottom.

Water Soldiers grow in still and slow-flowing water, pools and oxbow lakes where they sometimes form thick, spreading and impenetrable masses, thereby contributing to the silting-up of reservoirs and canals. Nevertheless, they are relatively rare plants native to central Europe and western Siberia; they are locally distributed in western and north-western Europe. In many places, e.g. in Scotland and Ireland, their occurrence is believed to be secondary, i.e. they are not indigenous but were introduced there.

Water Soldier is a perennial herb with a short rhizome from which rises a dense, many-leaved rosette of rigid leaves that are three-sided in cross-section (1) and spiny-serrate, prickly on the margin. This characteristic is what gave the plant its name — for the Greek word *stratiótes* means armed soldier.

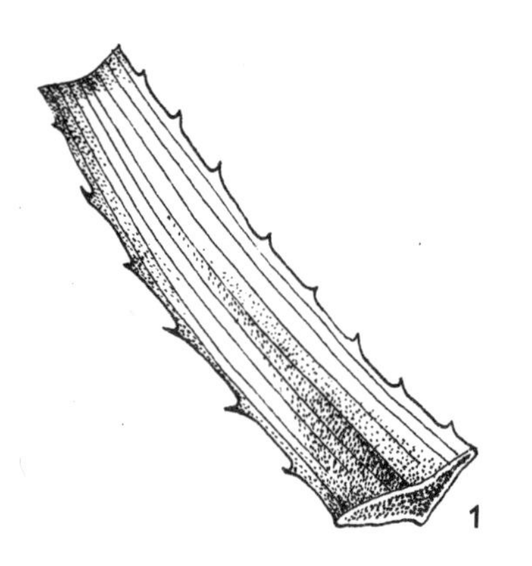

The surprisingly lovely, white, trimerous flowers are dioecious. The male flowers are long-stalked with numerous stamens, of which only about 12 have anthers. The female flowers are smaller with short stalks and six stigmas. The fruit is a spiny, leathery capsule. Water Soldiers, however, multiply readily by vegetative means, by means of new, daughter leaf rosettes formed at the tips of the long creeping stolons growing from the base of the parent plant.

The flowering period is from June onward, often until the beginning of September.

Common Hornwort Ceratophyllaceae
Ceratophyllum demersum L.

The delicate thread-leaved stems of hornwort are not in the least reminiscent of water-lilies at first glance but the hornwort family is classed in the order of water-lilies (Nympheales) — even though as the most distant type. The tissues of hornworts are very simplified: the conductive tissues consist of a single vessel, a rigid 'canal' in the centre of the stem surrounded by phloem. The structure of the leaves is different too — forked once or twice, they are arranged on the stem in whorls.

The hornwort family consists of a single genus — *Ceratophyllum* — with three species native to the southern parts of the USSR, one indigenous to the Far East and two that are nearly cosmopolitan. Hornwort has existed on the territory embracing present-day Europe for the past 70 million years or so — from the middle of the Tertiary period.

Ceratophyllum demersum grows in still and slow-flowing water, in pools, oxbow lakes and ponds — particularly in lowland districts. It does well also in deeper water where it finds the light conditions more congenial (diffused light); sometimes it forms thick, continuous, tangled masses that are a troublesome obstacle to fishing with nets. Neither this nor the Spineless Hornwort (*C. submersum* L.) are particularly suitable for aquariums, besides which it dies within a single season after being moved from its natural habitat. The less common, thermophilous Spineless Hornwort, however, tolerates brackish water and replaces Common Hornwort in such places.

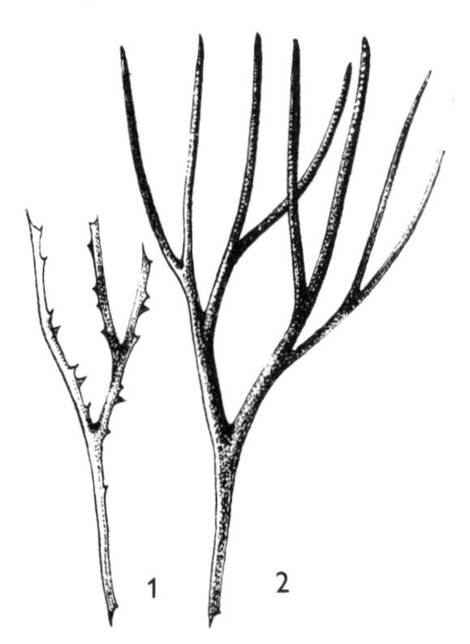

Hornworts are perennial aquatics that cannot exist in a different environment, one without water. Their roots are rudimentary or absent, or else replaced by short 'rhizoids'. The leaves are regularly forked — in *C. demersum* once or twice (1), in *C. submersum* three times (2) — into thread-like segments and arranged in whorls on the stem. The stems are branched and may be 1—2 m long.

C. demersum rarely produces flowers but when it does they are unisexual, stalkless and grow singly from the axils of the leaves. Male flowers have 10 to 20 stamens and twelve perianth segments; female flowers have approximately ten perianth segments and only a single ovary that develops into a nut with three spines (3).

If the plant flowers the flowers are produced from June to September.

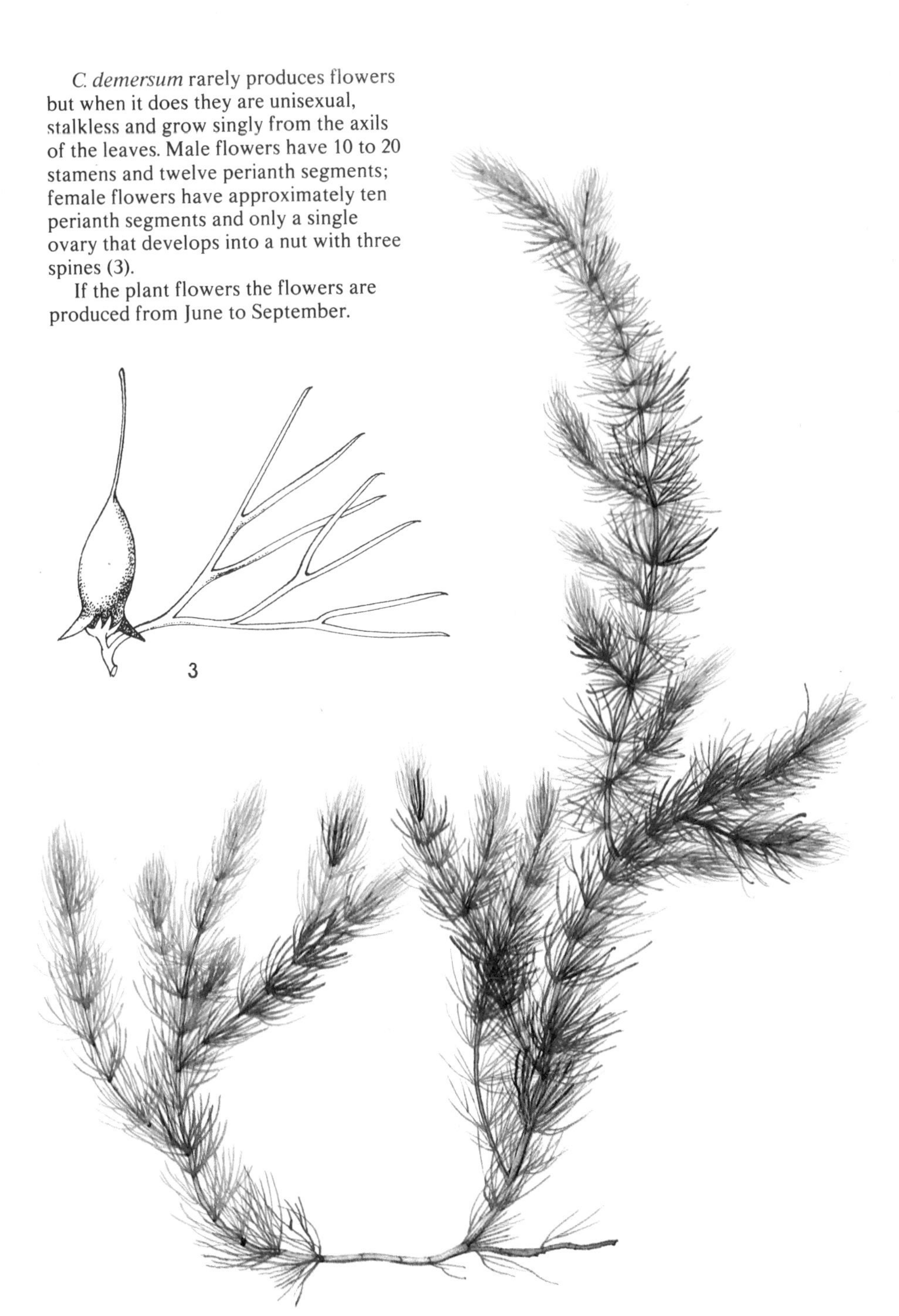

Spiked Water-milfoil
Myriophyllum spicatum L.

Haloragaceae

Comparison of water-milfoils and hornworts serves as a graphic illustration of how the same environment influenced the evolution and appearance of plants that are systematically quite distant.

The Haloragaceae are a family of aquatic and bog plants distributed from the tropics to the temperate regions, chiefly in the southern hemisphere and on the Australian coast. The genus *Myriophyllum* includes approximately 40 species. Australian species are generally amphibious, American species mostly aquatic. Spiked Water-milfoil is a practically cosmopolitan plant found in the waters of Europe, North America and north Africa; it is absent in South America and Australia. Its lovely, decorative stems may be up to 2 m long and the leaves almost 2.5 cm long. It grows scattered in still and slow-flowing water either singly or in larger masses. Unlike many other aquatic plants that develop long drifting stems or leaves in water currents water-milfoils develop only short stems, barely 30 — 50 cm long, in rapidly-flowing water. Spiked Water-milfoil also does well in deep water, in clean lakes as well as in calcareous water. The Alternate-flowered Water-milfoil (*M. alterniflorum* DC.), on the other hand — a species more common in west oceanic Europe and on the east coast of North America — requires clean, cool, non-calcareous, oxygen-rich water with gravelly bottom and a depth of up to 1 m.

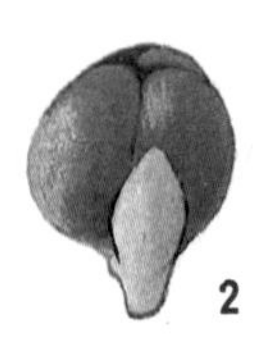

The scientific name derived from the Greek words *myrios*, meaning innumerable or very many, and *fyllón*, meaning leaf, aptly describes the appearance of these plants.

Water-milfoils are perennial aquatics with creeping rhizomes. The leaves are usually arranged in five-leaved (*M. verticillatum* L.) or four-leaved (*M. spicatum*, *M. alterniflorum*) whorls; they are finely pinnatisect with 6 to 20 thread-like segments on either side of the leaf (1). The stems of Spiked Water-milfoil are conspicuously reddish, which gives the 'floating' plant an unreal, exotic look. The flowers are borne in

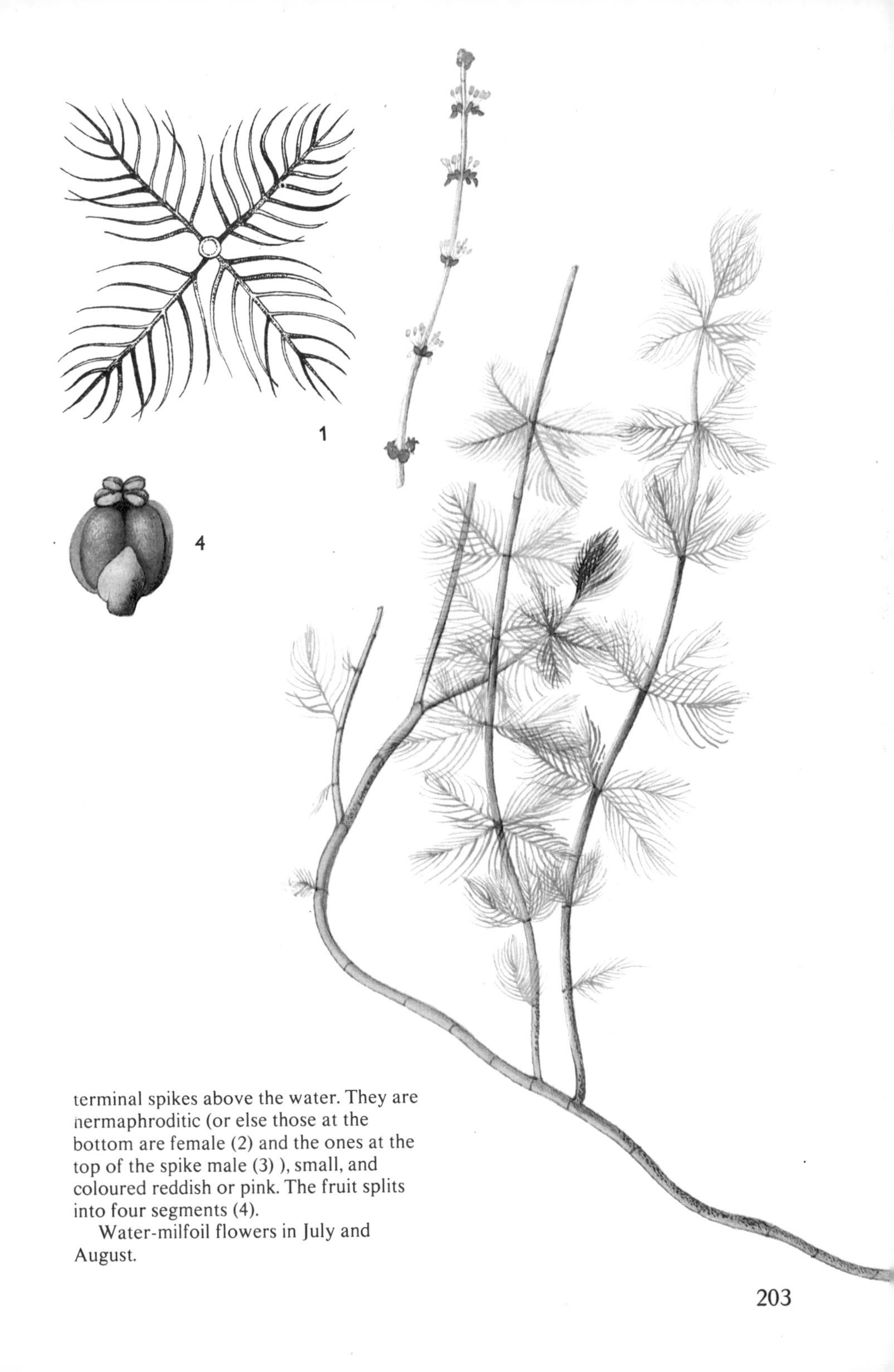

terminal spikes above the water. They are hermaphroditic (or else those at the bottom are female (2) and the ones at the top of the spike male (3)), small, and coloured reddish or pink. The fruit splits into four segments (4).

Water-milfoil flowers in July and August.

Fringed Water-lily
Nymphoides peltata (Gmel.) Kuntze

Menyanthaceae

Unlike the related gentians, often found high up in the mountains and far north, most members of the genus Nymphoides are thermophilous plants of tropical regions. Even the Fringed Water-lily is slightly thermophilous. It forms two kinds of jointed shoots: long shoots (some more than 1 m) with 8—20-cm-long internodes in summer and short shoots, barely 2—20 cm long, in autumn when the temperature of the water drops. It is the former, the long summer shoots that contribute to the plant's rapid spread. The leaves are likewise of different kinds, depending on the light and temperature. The first floating leaves appear as early as late April and are pale green. The later, definitive leaves are rigid, dark green above, reddish-violet beneath and finely dotted. The final, autumn leaves contain even more anthocyanins (pigments producing blue to red colouring) and are dark greenish-violet. Besides floating leaves, plants growing in deep water also produce smaller submerged leaves; on the exposed bottoms of ponds drained for the summer they may occur as terrestrial forms with short-stalked leaves.

The Fringed Water-lily is a Eurasian species with a subcontinental distribution extending from central Europe (as far north as Holland) to China; however, it was grown as an ornamental in many places and appears to have become locally established.

The Fringed Water-lily slightly resembles the White Water-lily. It is a perennial plant with creeping rhizome (1) bearing rounded stems. The leaves are nearly opposite and have long stalks in sheaths. The striking flowers are long-stalked and five-lobed; the golden-yellow corolla is funnel-shaped, up to 3 cm in diameter, and the lobes are ciliate (compare with the flowers of the related Bogbean on page 116). The flowers are usually pollinated by bees and have been found to be heterostylous, i.e. having styles of different lengths. The fruit is a pointed capsule.

The flowering period is in July and August.

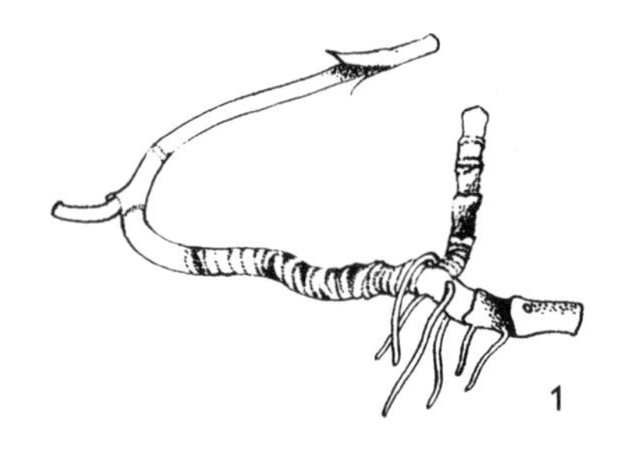

Common Duckweed
Lemna minor L.

All textbooks on botany usually end with duckweeds and duckweeds are also one of the last families in the current classification of higher plants. They are minute floating aquatics with organs reduced to a minimum size. No other group of angiosperms exhibits such reduction. The plant body consists of a leaf-like organ believed to be a single ± modified true leaf or flat, leaf-like stem; in the genus *Spirodela* the flat (large) leaf may be a stem for it has two scale-like true leaves on the underside. Duckweeds rarely produce flowers but they multiply readily by vegetative means. In the parent 'leaf' there is a small cavity containing a growing point which develops into a new segment, that in most duckweeds detaches itself but in some remains on the plant (Ivy Duckweed – *Lemna trisulca* L.).

Duckweeds are practically cosmopolitan, being readily dispersed even great distances, for instance on the feet and feathers of water birds. They are found on the surface of almost every calm body of water, often in large masses. The concentration of the various separate species in the community may change and fluctuate during a single growing season. For example: on one pond in a single year Great Duckweed *(Spirodela polyrhiza)* increased in number from May to July and then its numbers declined. Precisely the opposite was true in the case of Ivy Duckweed, whereas Common Duckweed, upon reaching its spring peak, continually declined in number until autumn.

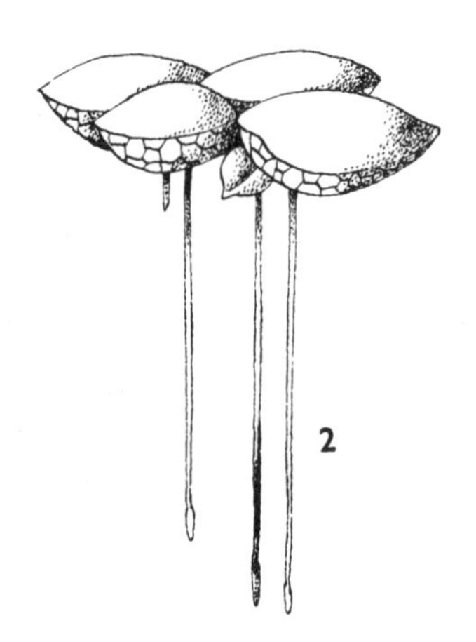

Common Duckweed (*Lemna minor* L.) (1) has flat, floating leaves (segments) coloured green on both sides and a single root. Gibbous Duckweed (*Lemna gibba* L.) has leaves strongly swollen beneath (2). Ivy Duckweed (*Lemna trisulca* L.) has submerged, ovate-lanceolate leaves tapering at the base into a long stalk and joined together after several 'generations' (3). Great Duckweed (*Spirodela polyrhiza* (L.) Schl.) has rounded-ovate leaves, conspicuously reddish on the underside, with tuft of thread-like roots (4).

When duckweeds do flower, the flowers are minute and consist solely of a single stamen or single pistil.

1
4
3

FLOWERS – STRIKING AND INSIGNIFICANT

Some meadow flowers are very showy and make a striking display with their profusion of colours. Others, however, are less showy and even inconspicuous.

The narcissus is a good example of the first group. The flower has the classic structure consisting of a well-developed perianth and regularly formed reproductive organs (stamens, pistil) and is sufficiently large to attract insect pollinators. Other plants, usually ones with small flowers, solved the problem of attracting insects by bearing the blossoms in clusters, in many-flowered inflorescences. In many instances this serves a twofold purpose. By themselves the individual flowers might be overlooked by insects but the colourful patch made by the whole cluster is a different matter. And secondly, the flowers in most plants open in succession (only in a few do the flowers that make up the inflorescence open simultaneously), thereby prolonging the period of possible pollination. The manner in which they open is the characteristic of certain groups of plants and of certain types of inflorescence.

The greatest perfection in the present stage of plant evolution has probably been achieved by those of the composite (Compositae) family. The lovely, typical meadow 'flowers' of daisies are in fact multi-flowered inflorescences composed of a great many central tubular flowers (disc florets) and numerous marginal ligulate flowers (ray florets). The whole appears to be a single flower but in reality it is an inflorescence.

The first and foremost striking aspect of the meadow flower is the

The large and colourful flower of Narcissus is a natural visual bait for insect pollinators.

The 'flower' of the Ox-eye Daisy is actually a complex inflorescence composed of numerous tiny flowers. However, because it fills its function as visual bait for insects so admirably it is also called a 'biological flower'.

colour — a veritable kaleidoscope — not the size. And as the various species do not bloom at the same time such a meadow presents an everchanging picture. These colour effects depend primarily on the concentration of the various species in the meadow community and naturally also on whether the plants of a given species all flower at the same time.

Large-flowered plants and ones with brightly coloured blossoms are the ones people notice most in meadows, even though they are in the minority. Most numerous are the grasses (Gramineae) and grass-like plants. Their flowers, however, are very insignificant — minute, as a rule — and even though they are arranged in large inflorescences they remain hidden from public view. Perhaps also because their colours are relatively nondescript. They are generally greenish with only an occasional colour accent in the form of violet or shining yellow anthers.

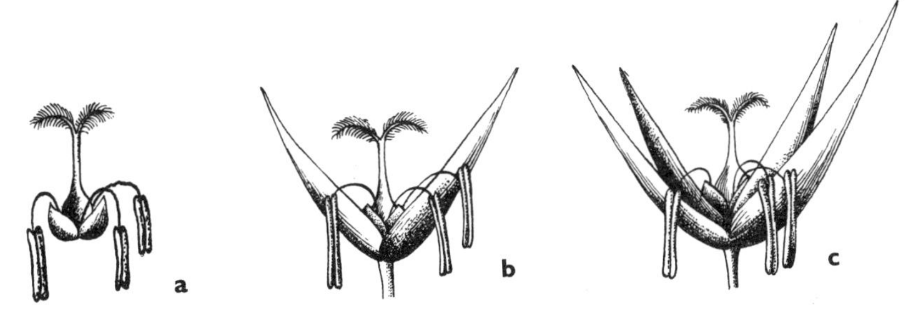

To better know the structure of the flower of grasses we must examine it one part after the other. In the centre of fig. a. is a pistil with two feathery stigmas, three stamens with long filaments and lodicules. The reproductive organs proper are protected by an outer lemma and inner palea (fig. b). The entire one-flowered spikelet is usually enclosed further by two glumes — one above and one below — (fig. c).

The flowers of grasses differ from the flowers of other plants — their structure is unique. The pistil was originally composed of three carpels. In existing grasses it appears to be a single unit without any trace of the fusion of carpels on the ovary. The short styles (usually two) are terminated by prominent feathery stigmas. The stamens have very thin filaments attached to the anther by means of a movable joint. In some grasses the weight of the pollen sacs causes them to protrude and seemingly hang from the flower when it is in full bloom, thus being swayed readily by the breeze and releasing the pollen grains. Grasses usually have three stamens. Originally there were six arranged alternately in two rings of three (still true of rice), but in the course of evolution those forming the inner ring atrophied and in most grasses only the three in the outer ring remained.

The flowers of grasses are seemingly without a perianth. However, the male and female reproductive organs are concealed between two enclosing bracts — an outer (lower) bract termed the lemma and an inner (upper) bract termed the palea. The lemma is a rounded, convex bract with a single rib (keel) often terminated by an awn. The palea has two prominent ribs, two terminal teeth and is awnless. The Czech botanist L. Čelakovský determined that the palea probably developed from the fusion of the two original perianth segments and the lemma from a subtending bract.

A further 'non-traditional' feature of the flowers of grasses are the lodicules. These are minute outgrowths at the base of the flower that swell up at flowering thereby opening the otherwise tightly-shut flower by separating the lemma and palea and assisting the escape of the anthers and feathery stigmas. The lodicules, usually two (very

Before grasses flower, the multiflowered spikelets are tightly closed (left); in full bloom the separate flowers are clearly visible (right).

The flowers of rushes of the genus *Juncus* measure only a few millimetres and by themselves would surely pass unnoticed — even though their structure is not different from that of tulips. The top picture shows the closed flower and the bottom the open flower of Hard Rush *(Juncus inflexus).*

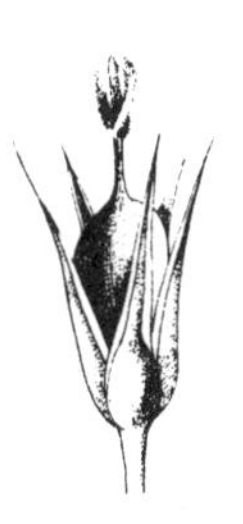

occasionally three), are remnants of the former inner ring of perianth segments.

The flowers of grasses are mainly arranged in two rows in spikelets enclosed at the base by glumes. The spikelets (one-flowered or multi-flowered) may be arranged in various ways, forming various types of inflorescences — spikes, racemes or panicles (e.g. meadow-grass, oat-grass, brome). These inflorescences may sometimes be contracted — dense and spike-like (timothy and foxtail grass).

All grasses are anemophilous, i.e. pollinated by the wind. The pollen carried by the wind readily catches in the feathery stigmas. Grasses produce an enormous amount of pollen, as do all wind-pollinated plants. When they are in flower the pollen in the air causes allergic reactions in some people, e.g. hay fever. In some grasses various parts of the flower are covered with long hairs which are very useful in the dispersal of the fruit, serving as a sort of 'flying apparatus'.

The fruit of grasses, called a grain, is rich in starch. The starch grains of each group of grasses differ in size and shape and are characteristic for each group of grasses. They are thus an important means of identification not only for purposes of classification (plant taxonomy) but also in forensic medicine and in the determination of quality, for many species of grasses are basic items of man's diet.

Grasses are truly a major component of meadow communities. The closer they are to water, however, the greater the number of other plants in their midst — plants that resemble grasses. These are first and foremost members of the sedge (Cyperaceae) and rush (Juncaceae) families, which likewise have relatively small, insignificant flowers.

The flowers of rushes are hermaphroditic and regular with well-developed perianth composed of an outer and inner ring of six scale-like or bract-like segments — three in each ring. The stamens are likewise six in number in two rows of three and the ovary is trilocular. Apart from size and colour the flower is thus no different from that of tulips, for instance.

Plants of the order Cyperales are believed to be derived from the order Juncales by some botanists. Their flowers are greatly reduced and even more greatly adapted for pollination by the wind. The perianth is sometimes completely absent or modified into feathery growths (for instance in the bulrush). Species with unisexual flowers far outnumber those with hermaphroditic flowers. The inflorescence is usually a spike or raceme composed of individual spikelets; sometimes the inflorescence is contracted into a spike or head. The spikelets of male flowers are generally more slender and located at the top of the inflorescence (in the uppermost spikes) whereas the female flowers (and spikelets of female flowers) are thicker and usually located at the bottom of the inflorescence (in the bottom spikes). The male flowers generally have only three stamens; the pistils of the female flowers are usually composed of two to three fused carpels but the ovary is unilocular and contains only a single ovule. In sedges the female flower is enclosed by a bract called a perigynium that tapers into a beak from which the styles and stigmas protrude. The fruit is an achene enclosed by the perigynium so that the whole resembles a follicle — a one-seeded fruit.

Though sedges are very similar to grasses they differ markedly in many respects. From an evolutionary viewpoint, the two groups have nothing in common and whatever resemblance there is is mere convergence (i.e. the development of similarities in unrelated organisms living in the same environment).

Small, insignificant flowers are also produced by other plants of wetlands and water. The spadix of sweet-flag, for instance, is composed of as many as 800 minute flowers that in detail are practically no different from the normally developed, trimerous flowers of other monocotyledonous plants. In some plants, however, the organs have been reduced to the maximum degree. The flowers of duckweeds are rarely seen, not just because they are produced only occasionally but because they are so insignificant. They are unisexual with the male flowers reduced to a single stamen and the female flowers to a single pistil, developed from a single carpel, with a single style and stigma. They are formed in a minute cavity and enclosed by a tiny spathe-like sheath.

STALWART GRASSES AND GRASS-LIKE PLANTS

Some of the fanciful inventions of Jules Verne's novels have long since become commonplace things and have even been surpassed. However, one thing man will never achieve is the strength of a stem of grass. None of the existing materials or technical inventions can match the stem of a cereal grass in simplicity and parameters, for instance. In view of the amount of 'building material' used, the length (i.e. the height) of the 'structure' compared to its weight is simply impossible to be emulated by man-made materials at the present time. The base/height ratio of such a stem is roughly 1:100 to 1:300 whereas the ratio for similar man-made structures — chimneys and TV towers — is scarcely 1:10. A grass stem will withstand the force of any wind — even if it is 5 m tall, as is the case in some reeds. Structurally it is a hollow cylindrical beam. The botanist will tell you that it is a hollow, rounded stem with central cavity formed by the drying and breaking-up of the pith in the core; this is true of most grasses, apart from a few exceptions such as millet and maize. The epidermis or outer layer is made up of cells with cutinized membranes, often permeated with silicon dioxide which further strengthens the stem. Beneath the outer layer there are two rings of vascular bundles with bands of tissue composed of thick-walled, dead cells that likewise strengthen the stem.

In most grasses the hollow stem is jointed and closed at the joints or nodes by solid tissue. Though strengthening the stem to some de-

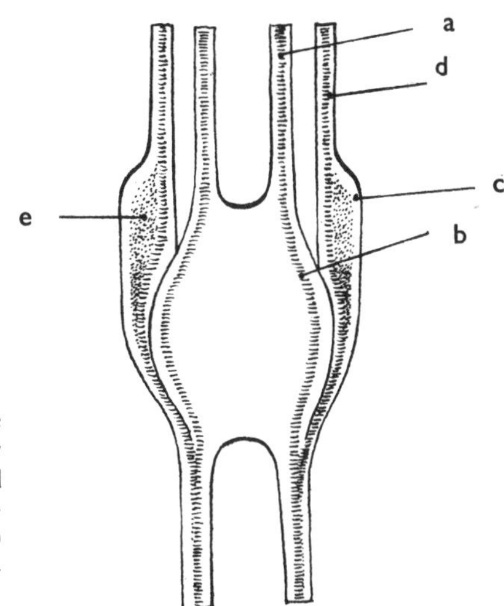

Nodes are an important feature of the stems (culms) of grasses. The hollow stem (a) is swollen at the node (b) and contains meristematic tissue. Also swollen (c) is the enveloping leaf sheath (d) which at the point of the swelling contains strengthening tissue (e).

gree their prime function is as mechanisms of growth. The grass stem may withstand the force of the wind but not damage caused by flood waters, hail, and the like; firm as it may be, even such a stem will bend, be torn up or laid flat. Grasses, however, are light-loving plants and in a mass that has been flattened to the ground they would shade one another. In such a case, a geotropical reaction (i.e. response to the earth's gravity) sets in, stimulating the growth of meristematic tissues located in the nodes. In such a prostrate stem the meristematic cells on the underside of the node immediately begin to grow and multiply rapidly and the stem slowly begins to straighten again at the node.

Grasses, as a rule, are plants with high ground coverage; that is why they are the prevailing plants in meadow communities. Besides other reasons this is due to their ability to produce tillers (shoots that spring from the base of the stalk and form many-stemmed tufts) as well as of spreading rapidly by vegetative means. Every gardener knows what the underground stolons of couch grass can do in a flower bed. Reeds, the largest of the European grasses, are noted for their rapid coverage by vegetative propagation. Though reeds have already been mentioned in the introduction let us provide a few more statistics: the underground rhizomes of a single plant increase 2 to 3 m in length in one year; a three-year colony formed by the spread

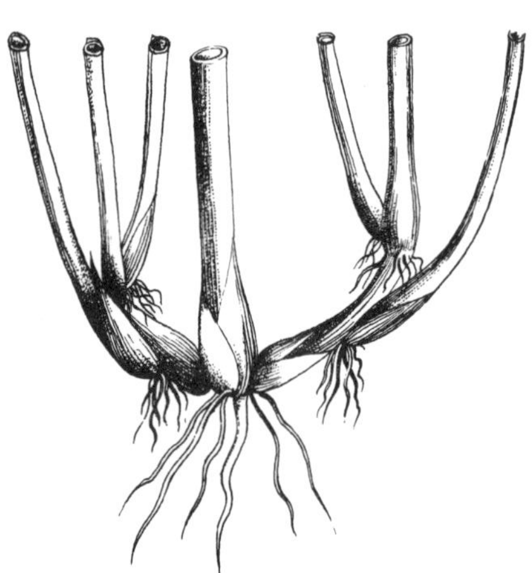

A single grain (seed) may give rise to several culms with spikes, for grasses are distinguished by the ability to produce tillers (shoots that grow from the base) and, besides the thick main shoot, successively produce shoots of the first and second order.

Rapid spread by vegetative means is a distinguishing feature of so-called creeping grasses; underground — in some grasses also on the surface — runners continually give rise to new 'plants'.

Due to the rapid formation and growth of its rhizomes a single reedmace can cover an area of several square metres within a single growing season.

of a single individual covers an area of nearly 14 — 20 sq m and the length of its rhizomes may be as much as 30 m! It spreads and binds the soil about 20 — 30 cm below the surface but may penetrate to depths of 50 — 100 cm and retains its vigour for as long as six years. This likewise testifies to the sturdiness and immense vitality of grasses — those important plants of meadows and wetlands.

Of the marsh grasses reedmace is particularly noteworthy for the rapidity of its spread by vegetative means.

In an older mass of Great Reedmace there are from three to seven fertile female inflorescences (spadixes) to a square metre. Each such inflorescence produces approximately 220,000 downy achenes distributed great distances by the wind. Within one year a single germinated seed forms a colony measuring about 10 sq m and within two years it covers an area of up to 50 sq m.

The rate at which reedmace spreads (slower at the beginning of the growing season but rapidly in July and August) is determined by the morphological characteristics of the root system, particularly its form of branching. During the growth period the lateral buds at the base of the surface shoot (above ground) develop into rhizomes about 0.5 to 1 m long that spread horizontally in the surface layers of mud. They do not branch but bear new shoots at the tip, likewise with lateral buds at the base that shortly begin growth, thereby perpetuating the process. For a certain time the newly-formed surface and underground parts of the reedmace remain joined and form a greatly-branched colony. After about three years the parent plant dies and the detached 'offspring' continue in the old tradition. Each such great reedmace colony can produce as many as three new surface shoots in a single day, during the period of rapid growth; in that same period up to 1 m of new rhizomes are produced daily by each colony. These numbers, of course, apply only to the first year of growth. In the second year up to 10 new shoots are produced every day and nearly

Aerenchymatous tissue and large air-filled intercellular spaces may be seen also on a section of the rhizome of the Common Reed.

4 m of rhizomes are added daily by each colony. In congenial conditions a single seedling may produce 200 new tillers and approximately 120 m of rhizomes during its first year and about 1,000 tillers and 360 m of rhizomes during its second year.

Mutiplying as they do, however, reedmaces have provided man with an enormous quantity of raw material which in districts with many ponds and lakes was often the natives' only means of livelihood. Nowadays this folk industry — the making of various items such as bags, slippers, and hats — is a more or less nostalgic reminder of bygone times. Generally articles are only made for the tourist trade and not for daily use by rural folk.

BODIES ADAPTED TO LIFE IN WATER

It is often said that there is no life without water. Applied to aquatic plants, however, this saying would have to read 'there would be no life without air', for even their submerged parts cannot do without oxygen. Aquatic plants have a well-developed system of 'gas pipelines' — intercellular spaces and ducts in their stems, roots and leaves. This serves not only for the exchange of gases but also reduces the specific weight thereby enabling them to float in water. Plant tissues with large intercellular spaces are called aerenchymatous tissues. In some plants they are incorporated in the structure of the organs and are not conspicuous on the exterior — they may be seen only if we cut the stem, leaf-stalk or root. In other species, however, these tissues form swollen or inflated vescicle ('floats'), for example in the bladder-like leaf-stalks of Water Chestnut (*Trapa natans*) or in the swollen leaf-stalks of the tropical Water Hyacinth (*Eichhornia crassipes*).

The roots of aquatic plants likewise have a different function to those of terrestrial plants. As a rule they serve only to anchor the plant in the bottom. In some aquatics the roots contain green colouring matter (chlorophyll) and carry on normal photosynthesis. Many aquatic plants have no roots whatsoever (e.g. bladderwort — *Utricu-*

laria) or else only greatly reduced ones as the single root of Common Duckweed — (*Lemna minor*).

The leaves of submerged aquatics have a simplified anatomy. They consist essentially of simple soft tissue made up of thin-walled undifferentiated cells containing chlorophyll and enclosed by an imperceptible epidermis. The vascular bundles are likewise usually reduced and there are no pores (stomata). In plants with floating leaves (White Water-lily, Yellow Water-lily, pondweed) the upper surface is 'coated' with a waxy layer and water rapidly runs off. All the pores are located on this upper surface which is exposed to the air above the water. If such leaves have pores on the undersurface, then they are usually never open. This is unlike most terrestrial plants which usually have pores on the underside of the leaf. For example, for every square millimetre of Elderberry, there are no pores above and 48 on the underside; Common Lungwort none above and 240 on the undersurface; Ribwort 165 above and 160 on the underside. However, Duckweed has 82 above and no pores beneath; Broad-leaved Pondweed 189 above and none beneath. In aquatics that have submerged or floating leaves and emersed aerial leaves the ratio is 115:0 (floating leaf of Arrowhead) and 112:25 (emersed leaf of Arrowhead); in the floating leaf of the aquatic form of Amphibious Bistort it is 120:0 but in the terrestrial form the displacement of the pores is practically no different from that of normal land plants — only 24 per sq mm above and 128 per sq mm on the undersurface!

A typical characteristic of many aquatic plants is growing leaves of different forms on the same stem (heterophylly): submerged leaves and emersed leaves (either rising above the water or floating on the surface).

Floating leaves are exposed to possible damage by the movement of the surface water (waves as well as ripples) and are therefore usually more rigid and leathery (caused by thickening of the cell walls), the better to withstand pounding by water.

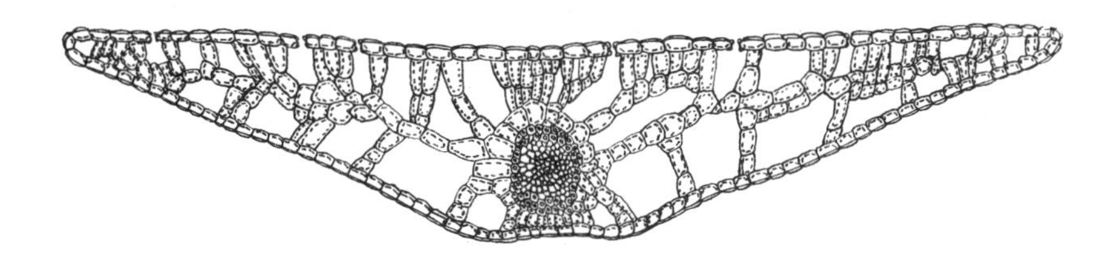

The floating leaves of pondweed have pores only on the upper surface and large air-filled intercellular spaces that buoy up the leaves.

Submerged leaves often have finely dissected leaf blades. This is an example of classic convergence for they have nothing in common either systematically or evolutionally with similar-looking plants *(Hottonia, Ceratophyllum, Myriophyllum, Batrachium).* Some plants, particularly in flowing water, grow floating leaves that are often very long and ribbon-like (Floating Sweet-grass — *Glyceria fluitans*). The purpose of such adaptations to the aquatic environment is to enlarge the leaf area in contact with the water and thereby its absorption capacity. Besides this they have proved to be very advantageous from the evolutionary viewpoint, well resistant to flowing water.

The aquatic environment appears to be universal and not as readily affected by climatic changes during the course of the day and year as dry land habitats.

Unusual and unique in this respect are the conditions on riversides and in riverbeds. Water courses are natural 'migratory routes' for many organisms — from lower plants to man. Rivers are the routes along which mountain elements spread to the lowlands and vice versa. The natural, unregulated channel is the most changeable element in the landscape. The environmental conditions there change not only from year to year but often even within a period of days during the passage of flood water, floating ice, and the like. Shoreline vegetation is thus in perpetual motion, its development is very dynamic and the succession of shoreline colonization is practically uninterrupted.

Potamogeton natans

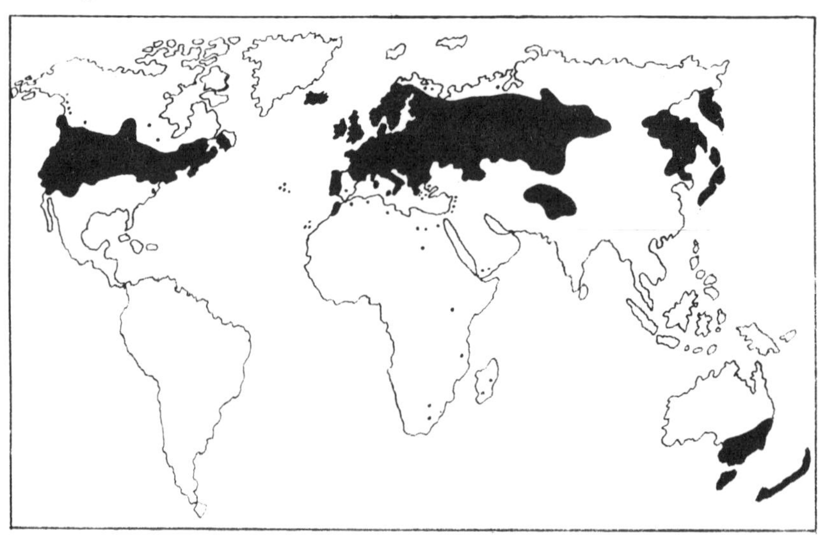

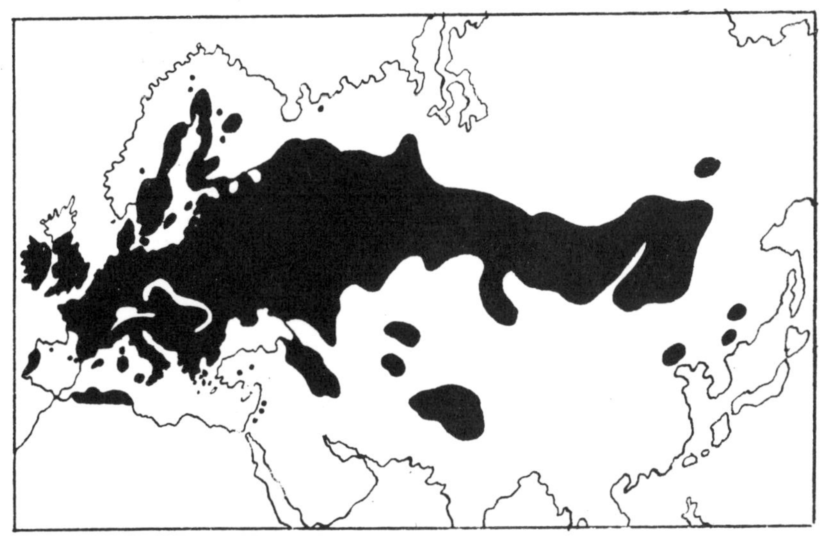

Butomus umbellatus

A river is an important means of transport for many plants; seeds as well as parts of plants may be carried great distances. The spring thaw and floating ice often damage (uproot) even waterside thickets and the plants are carried far from their original location by the current. That, for example, is how *Rosa majalis* Herrm. spread along several European rivers, as did whole colonies of Sweet Flag (*Acorus calamus*). Some adventive shoreline plants, particularly if they produce large numbers of viable seeds (Stick-tight – *Bidens frondosa*), likewise spread via the network of water courses. The spread of many plants in the opposite direction, upriver, is aided by animals (water birds and mammals) and sometimes by man himself.

Stratiotes aloides

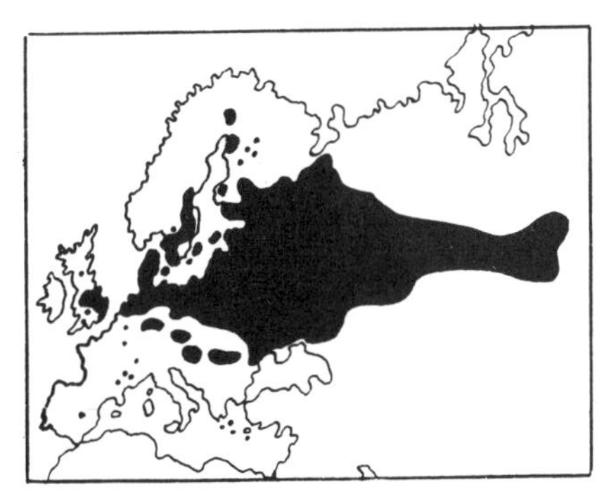

Holcus lanatus

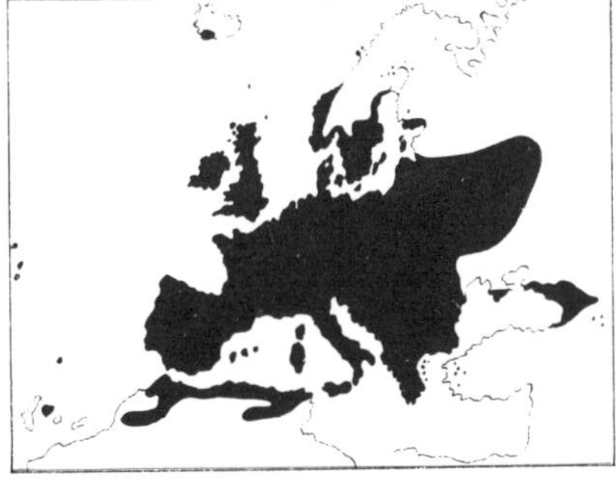

Isoëtes lacustris

AQUATIC AND SHORELINE PLANTS ON THE EARTH

Many aquatic and shoreline plants have a practically world-wide distribution. They are at home everywhere and thrive in practically all places. Examples are the oft-mentioned Common Reed *(Phragmites australis)* and Broad-leaved Pondweed *(Potamogeton natans).*

The second group are plants with a smaller, less extensive range from the viewpoint of distribution. They are not rare endemics but plants distributed more or less on a single continent. As shown on the separate maps they are mostly Eurasian or Euro-Siberian varieties.

Quite unusual in this context is one of the rarest elements of Europe's existing flora — the Common Quillwort. This relic of the flora of long-ago geological periods is now found only in a few glacial lakes in central Europe (the dotted section on map p. 220) and somewhat more frequently in north-western Europe. Its requirements of the water (purity, chemical composition and temperature) are factors that greatly limit its present distribution.

Meadow and aquatic plants rarely occur singly. They are social organisms and as a rule always occur in masses of characteristic appearance and texture. To show this, of course, is beyond the scope of this book and that is why it has tried all the more to acquaint the reader with the common as well as rarer plants of these places.

GLOSSARY

Achene a dry single seeded, non splitting fruit formed from one carpel.
Angiosperm flowering plant with the ovules borne within a closed cavity. The ovary becomes a fruit enclosing one or more seeds.
Annual a plant which flowers and dies within one year.
Anther the part of the stamen containing the pollen.
Anthocyanins water soluble colouring matters of plants.
Appressed pressed close to another organ but not united with it.
Awn a stiff bristle-like projection from the tip or back of the lemma in grasses.
Axil the angle between a leaf and a stem, often containing an axillary bud or flower.

Biennial a plant which takes two years to complete its life cycle, producing leaves in the first year, flowering and then dying in the second.
Bifid split deeply in two.
Bract a small leaf, green or scale-like, either at the base of a flower stalk or grouped beneath a flowerhead.
Bracteole leaf, generally very small, borne on the stalk of a flower.
Bulb an enlarged, underground bud with fleshy scale leaves.
Bulbil a small bud or tuber-like structure formed at the base of a leaf or in place of a flower, which breaks off and grows into a new plant.

Calyx a collective name for the sepals either separate or joined.
Capsule a dry fruit which splits open releasing the seeds.
Carpel one female reproductive unit of a flower, consisting of an ovary, style and stigma. The carpels may be separate or fused together.
Chlorophyll a green pigment found in the cells of algae and higher plants, which is fundamental in the use of light energy in photosynthesis.
Chloroplast a small body containing chlorophyll, situated in the protoplasm of a plant cell.
Conjugation the union of gametes in sexual reproduction.
Corolla a collective name for the petals either separate or joined.
Culm stem of a grass or sedge.
Cuticle a superficial non-cellular layer that covers the tissues and in many cases cuts down evaporation of water.
Cyme an inflorescence in which the main axis ends in a flower.

Deflexed bent sharply downwards.
Dehiscent description of plant structures that split open to shed seeds and spores.
Dichasium a cyme in which the branches are opposite and approximately equal; adj. **dichasial.**
Dicotyledon a plant of the major group which is characterised by having two seed leaves, flower parts mostly in fours or fives and usually broad, net-veined leaves.
Dioecious plants those with male and female flowers on separate plants.
Drupe a fleshy fruit containing one large seed which is surrounded by a stony layer.

Endemic occurring naturally, not introduced. Usually referring to a species unique to a single country or district.
Endosperm the tissue that surrounds and nourishes the embryo in seed plants.
Eutrophic rich in nutrients.

Floret a small flower which forms part of a compound head as in the family Compositae, or in a compact or open inflorescence as in sedges and grasses.
Flower a specialised reproductive shoot, usually consisting of sepals, petals, stamens and ovary.
Follicle a dry dehiscent fruit formed of one carpel.

Gamete a sex cell that fuses with another gamete in sexual reproduction to produce a zygote. The zygote develops into a new plant.
Genus a term of classification for a group of closely related species. It is the first of the two Latin names normally used to describe an organism.
Geotropism orientation of the plant under the influence of gravity so that the shoots grow vertically upward and the roots downward.
Glabrous without hairs.
Glumes a pair of dry scales enclosing the base of a grass spikelet.
Gymnosperms primitive seed plants with the ovules unprotected and not enclosed within an ovary. The class Gymnospermae includes conifers.

Halophilous freshwater plant capable of surviving in salt water.
Herbaceous plants non-woody, soft and leafy plants with parts that do not persist above the ground year after year.
Hermaphrodite having both male and female parts.
Heterostylous the length of the style in relation to the other parts of the flower differing in the flowers of different plants.
Hybrid a plant produced from cross-breeding of two different species and processing characters from both parents.
Hydrophytes plants that inhabit water or very wet places.
Hygroscopic able to absorb water, showing a change of form as a result.

Indumentum the hairy covering as a whole.
Inflorescence the flowers and associated structures on a flower stem; also the pattern of branching of the flower stem.
Involucre a group of bracts or other leafy structures developing below the flowers.

Laciniate deeply and irregularly divided into narrow segments.
Lactiferous containing latex.
Latex a milky juice, white or coloured.
Leaflet a separate segment of a leaf which often resembles a leaf but has no associated bud.
Lemma the lower bract that encloses the individual flower of a grass.
Ligule a small membrane at the junction of the leaf-sheath and the base of the leaf-blade in grasses.
Littoral zone part of seashore inhabited by plants, below average water level.
Lobed divided, but not into separate parts.
Lyrate shaped like a lyre.

Meristem a region of active cell division, which gives rise to the permanent tissues of the plant. Apical meristems are the growing tips of shoots or roots.
-merous e. g. in 5-merous (= pentamerous); having the parts in fives.
Mixotrophic containing two or more methods of nutrition.
Monochasium a cyme in which the branches are spirally arranged or alternate or one is more strongly developed than the other; adj. **monochasial.**

Monocotyledon a plant of the major group which is characterised by having one seed leaf, flower parts mostly in threes, and usually narrow leaves with parallel veins.
Monoecious plants those with separate male and female flowers on the same plant.

Nectary a glandular body within the flower, varying in shape and position, which secretes honey-like nectar.
Niche the special position occupied by an organism in a particular habitat, including its relationship with other organisms.
Node the position on a stem where one or more leaves arise.

Orbicular rounded, with length and breadth about the same.
Opposite of two organs arising at the same level on opposite sides of the stem.
Ovary the central part of the flower containing the ovules which later develop into seeds.

Palmate having leaflets radiating from the same point.
Panicle strictly a branched racemose inflorescence though often applied to any branched inflorescence.
Pappus a circle of hairs, bristles, or scales (representing the calyx) at the top of the ovary of flowers of the family Compositae.
Perennial a plant which lives for more than two years, usually flowering in each year.
Perianth a collective term for the sepals and petals that together form the asexual part of the flower.
Petal a segment of the corolla, often brightly coloured.
Photosynthesis a series of chemical reactions in the tissues of green plants that synthesize organic compounds from water and carbon dioxide using energy absorbed by chlorophyll from light.
Phototropism a growth response to light related to the direction of the stimulus. Shoots usually show positive phototropism, ie. grow towards the light, whereas roots usually show negative phototropism, ie. grow away from the light.
Pollen the microspores of conifers and flowering plants, each of which contains a male gamete.
Pollination the process by which pollen is transferred from the anther to the stigma.
Pinnate bearing leaflets along each side of the leaf axis.
Pistil the ovary of a flower with a style and stigma.
Plural-yearly more than twice a year.

Rachis main axis of the inflorescence.
Raceme an unbranched inflorescence in which the flowers are borne on the stalk of a single flower.
Relict a surviving organism characteristic of an earlier time.
Revolute rolled downwards.
Rhizome a creeping underground stem, often swollen with food reserves, from which grow leaves and stems each year.
Rosette a radiating cluster of leaves, usually lying close to the ground.

Scape the flowering stem of a plant all the foliage leaves of which are radial.
Schizocarp a syncarpous ovary which splits up into separate 1-seeded portions (mericarps) when mature.
Seed a structure within the ovary, produced by fertilization of the ovule and consisting of a seed coat, food reserves and an embryo capable of germination.
Sepal a segment of the calyx, usually small, green, and leaf-like.
Sessile without a stalk.
Somatic cell nonreproductive cell.
Spadix spike with a swollen axis, enclosed in a spathe.
Spathe large foliar organ which subtends a spadix.
Species a group of individuals with similar characteristics that can interbreed but do not normally breed with individuals from another species. The specific name is the second of the two Latin names used to describe an organism.
Spike a simple elongated inflorescence with sessile flowers occurring up the stem, the youngest at the top.
Spikelet a group of one or more florets in the grass family.
Spore a single-celled or multicellular asexual reproductive body that detaches from the parent plant and eventually grows into a new plant. Resting spores are capable of withstanding unfavourable conditions.
Spur a long, normally nectar-producing projection from a sepal or petal.
Stamen the male reproductive structure, usually stalked.
Staminode an infertile, often reduced, stamen.
Stigma the receptive part of the ovary, varying in form, on which the pollen germinates.
Stipule a lateral outgrowth, paired and varying in shape, which develops from the leaf base.
Stolon a creeping stem of short duration produced by a plant which has a central rosette or erect stem; when used without qualification is above ground; adj. **stoloniferous.**
Style a prolongation of the ovary which bears the stigma(s).

Tiller shoot of a plant which springs from the root or base of the original stalk.
Trifid split into three but not to the base.
Trifoliate with three leaflets, as in a clover leaf.
Tropism any growth response to a stimulus determined by the direction of the stimulus, as in phototropism.
Tuber a swollen, underground stem or root.
Turgid the swollen state of plant cells when they have taken up water to their fullest extent.
Turgor balance between the osmotic pressure of the cell sap and the elasticity of the cell wall.

Umbel an umbrella-shaped inflorescence with the flower stalks arising from a common point.

Vegetative propagation asexual reproduction in which part of a plant becomes detached and subsequently develops into a new plant.
Vein a strand of strengthening and conducting tissue running through a leaf or modified leaf.
Verticillaster inflorescence which looks like a dense whorl of flowers but is really a combination of two crowded dichasmial cymes.

Whorl a ring of three or more similar structures.

Xerophilous tolerant of a draughty habitat.

Zygomorphic description of flowers that are irregular and can be divided vertically in one plane only to produce identical halves.
Zygote the fertilized ovum before it divides to produce the cells of a new plant.

INDEX